THE CREATIVITY CODE

By the same author

The Music of the Primes:
Why an Unsolved Problem in Mathematics Matters

Finding Moonshine:
A Mathematician's Journey Through Symmetry

The Number Mysteries

What We Cannot Know:
Explorations at the Edge of Knowledge

THE
CREATIVITY
CODE

HOW AI IS LEARNING
TO WRITE, PAINT AND THINK

MARCUS DU SAUTOY

4TH ESTATE · LONDON

4th Estate
An imprint of HarperCollins*Publishers*
1 London Bridge Street
London SE1 9GF
www.4thEstate.co.uk

First published in Great Britain by 4th Estate in 2019

1 3 5 7 9 8 6 4 2

A catalogue record for this book is available from the British Library

HB ISBN 978-0-00-828815-0
TPB ISBN 978-0-00-829634-6

Diagrams redrawn by Martin Brown

Printed and bound in Great Britain by CPI Group (UK) Ltd, Croydon, CR0 4YY

MIX
Paper from
responsible sources
FSC™ C007454

To Shani,
for all her love and support,
creativity and intelligence

CONTENTS

1

THE LOVELACE
TEST

Works of art make rules;
rules do not make works of art.
Claude Debussy

The machine was a thing of beauty. Towers of gears with numbers on their teeth pinned to rods driven by a handle that you turned. The seventeen-year-old Ada Byron was transfixed as she cranked the handle of Charles Babbage's machine to watch it crunch numbers, calculate squares and cubes and even square roots. Byron had always had a fascination with machines, fanned by the tutors her mother had been happy to provide.

Studying Babbage's plans some years later for the Analytical Engine, it dawned on Ada, now married to the Earl of Lovelace, that this was more than just a number cruncher. She began to record what it might be capable of. 'The Analytical Engine does not occupy common ground with mere "calculating machines." It holds a position wholly its own, and the considerations it suggests are more interesting in their nature.'

Ada Lovelace's notes are now recognised as the first inroads into the creation of code. That kernel of an idea has blossomed into the artificial intelligence revolution that is sweeping the world today, fuelled by the work of pioneers like Alan Turing, Marvin Minsky and Donald Michie. Yet Lovelace was cautious as to how much any machine could achieve: 'It is desirable to guard against the possibility of exaggerated ideas that might arise as to the powers of the Analytical Engine. The Analytical Engine has no pretensions whatever to originate anything. It can do whatever we order it to perform.' Ultimately, she believed, it was limited: you couldn't get more out than you'd put in.

This idea was a mantra of computer science for many years. It is our shield against the fear that we will set in motion something we can't control. Some have suggested that to program a machine to be artificially intelligent, you would first have to understand human intelligence.

What is going on inside our heads remains a mystery, but in the last few years a new way of thinking about code has emerged: a shift from a top-down attitude to programming to a bottom-up effort to get the computer to chart its own path. It turns out you don't have to solve intelligence first. You can allow algorithms to roam the digital landscape and learn just as a child does. Today's code created by machine learning is making surprisingly insightful moves, spotting previously undiscovered features in medical images, and investing in shrewd trades on the stock market. This generation of coders believes it can finally prove Ada Lovelace wrong: that you can get more out than you programmed in.

Yet there is still one realm of human endeavour that we believe the machines will never be able to touch, and that is creativity. We have this extraordinary ability to imagine and innovate and to create works of art that elevate, expand and transform what it means to be human. These are the outpourings of what I call the human code.

This is code that we believe depends on being human because it is a reflection of what it means to be human. Mozart's requiem allows us to contemplate our own mortality. Witnessing a performance of *Othello* gives us the chance to navigate our emotional landscape of love and jealousy. A Rembrandt portrait seems to capture so much more than just what the sitter looks like. How can a machine ever hope to replace or even to compete with Mozart, Shakespeare or Rembrandt?

I should declare at the outset that my field of reference is dominated by the artistic output of the West. This is the art I know, this is the music I have been brought up on, the literature that dominates my reading. It would be fascinating to know if art from other cultures might be more amenable to being captured by the output of a machine, but my suspicion is that there is a universal challenge here that transcends cultural boundaries. And so although I make some apology for my Western-focused viewpoint, I think it will provide a suitable benchmark for the creativity of our digital rivals.

Of course, human creativity extends beyond the arts: the molecular gastronomy of the Michelin-star chef Heston Blumenthal; the football trickery of the Dutch striker Johan Cruyff; the curvaceous buildings of Zaha Hadid; the invention of the Rubik's cube by the Hungarian Ernö Rubik. Even the creation of code to make a game like Minecraft should be regarded as part of some of the great acts of human creativity.

More unexpectedly creativity is an important part of my own world of mathematics. One of the things that drives me to spend hours at my desk conjuring up equations and penning proofs is the allure of creating something new. My greatest moment of creativity, one that I go back to again and again, is the time I conceived of a new symmetrical object. No one knew this object was possible. But after years of hard work and a momentary flash of white-hot inspiration I wrote on my yellow notepad the blueprint for this novel shape. That sheer buzz of excitement is the allure of creativity.

But what do we really mean by this shape-shifting term? Those who have tried to pin it down usually circle around three ideas:

creativity is the drive to come up with something that is new and surprising and that has value.

It turns out it's easy to make something new. I can get my computer to churn out endless proposals for new symmetrical objects. It's the surprise and value that are more difficult to produce. In the case of my symmetrical creation, I was legitimately surprised by what I'd cooked up, and so were other mathematicians. No one was expecting the strange new connection I'd discovered between this symmetrical object and the unrelated subject of number theory. The fact that this object suggested a new way of understanding an area of mathematics that is full of unsolved problems is what gave it value.

We all get sucked into patterns of thought. We think we see how the story will evolve and then suddenly we are taken in a new direction. This element of surprise makes us take notice. It is probably why we get a rush when we encounter an act of creativity, either our own or someone else's.

But what gives something value? Is it simply a question of price? Does it have to be recognised by others? I might value a poem or a painting I've created but my conception of its value is unlikely to be shared more widely. A surprising novel with lots of plot twists could be of relatively little value. But a new and surprising approach to storytelling or architecture or music that begins to be adopted by others and that changes the way we see or experience things will generally be recognised as having value. This is what Kant refers to as 'exemplary originality', an original act that becomes an inspiration for others. This form of creativity has long been thought to be uniquely human.

And yet all of these expressions of creativity are at some level the products of neuronal and chemical activity. This is the human code that millions of years of evolution has honed inside our brains. As you begin to unpick the creative outpourings of the human species you start to see that there are rules at the heart of the creative process. Could our creativity be more algorithmic and rule-based than we might want to acknowledge?

The challenge of this book is to push the new AI to its limits to see whether it can match or even surpass the marvels of our human code. Can a machine paint, compose music or write a novel? It may not be able to compete with Mozart, Shakespeare or Picasso, but could it be as creative as our children when they write a story or paint a scene? By interacting with the art that moves us and understanding what distinguishes it from the mundane and bland, could a machine learn to be creative? Not only that, could it extend our own creativity and help us see opportunities we are missing?

Creativity is a slippery word that can be understood in many different ways in different circumstances. I will mostly focus on the challenge of creativity in the arts, but that does not mean this is the only sort of creativity possible. My daughters are being creative when they build their castles in Lego. My son is heralded as a creative midfielder when he leads his football team to victory. We can solve everyday problems creatively, and run organisations creatively. And, as I shall illustrate, mathematics is a much more creative subject than many recognise, a creativity that actually shares much in common with the creative arts.

The creative impulse is a key part of what distinguishes humans from other animals and yet we often let it stagnate inside us, falling into the trap of becoming slaves to our formulaic lives, to routine. Being creative requires a jolt to take us out of the smooth paths we carve out each day. That is where a machine might help: perhaps it could give us that jolt, throw up a new suggestion, stop us from simply repeating the same algorithm each day. The machines might ultimately help us, as humans, to behave less like machines.

You may ask why a mathematician is offering to take you on this journey. The simple answer is that AI, machine learning, algorithms and code are all mathematical at heart. If you want to understand how and why the algorithms that control modern life are doing what they do, you need to understand the mathematical rules that underpin them. If you don't, you will be pushed and pulled around by the machines.

AI is challenging us to the core as it reveals how many of the tasks humans engage in can be done equally well, if not better, by machines. But rather than focus on a future of driverless cars and computerised medicine, this book sets out to explore whether these algorithms can compete meaningfully with the power of the human code. Can computers be creative? What does it mean to be creative? How much of our emotional response to art is a product of our brains responding to pattern and structure? These are some of the things we will explore.

But this isn't just an interesting intellectual challenge. Just as the artistic output of humans allows us to get some insight into the complex human code that runs our brains, we will see how the art generated by computers provides a surprisingly powerful way to understand how the code is working. One of the challenges of code emerging in this bottom-up fashion is that the coders often don't really understand how the final code works. Why is it making that decision? The art it creates may provide a powerful lens through which to gain access to the subconscious decisions of the new code. And it may also reveal limitations and dangers that are inherent in creating code that we don't fully understand.

There is another, more personal, reason for wanting to go on this journey. I am going through a very existential crisis. I have found myself wondering, with the onslaught of new developments in AI, if the job of mathematician will still be available to humans in decades to come. Mathematics is a subject of numbers and logic. Isn't that what a computer does best?

Part of my defence against the computers knocking on the door of the department, wanting their place at the table, is that as much as mathematics is about numbers and logic, it is a highly creative subject, involving beauty and aesthetics. I want to argue in this book that the mathematics we share in our seminars and journals isn't just the result of humans cranking a mechanical handle. Intuition and artistic sensitivity are important qualities for making a good mathematician. Surely these are traits that can never be programmed into a machine. Or can they?

This is why, as a mathematician, I am attentive to how successful the new AI is being in gaining entry to the world's galleries, concert halls and publishing houses. The great German mathematician Karl Weierstrass once wrote: 'a mathematician that is not something of a poet will never be a true mathematician.' As Ada Lovelace perfectly encapsulates, you need a bit of Byron as much as Babbage. Although she thought machines were limited, Lovelace began to realise the potential of these machines of cogs and gears to express a more artistic side of its character:

> It might act upon other things besides number . . .
> supposing, for instance, that the fundamental relations
> of pitched sounds in the science of harmony and of
> musical composition were susceptible of such expression
> and adaptations, the engine might compose elaborate and
> scientific pieces of music of any degree of complexity
> or extent.

Yet she believed that any act of creativity would lie with the coder, not the machine. Is it possible to shift the weight of responsibility more towards the code? The current generation of coders believes it is.

At the dawn of AI, Alan Turing famously proposed a test to measure intelligence in a computer. I would now like to propose a new test: the Lovelace Test. To pass the Lovelace Test, an algorithm must originate a creative work of art such that the process is repeatable (i.e. it isn't the result of a hardware error) and yet the programmer is unable to explain how the algorithm produced its output. This is what we are challenging the machines to do: to come up with something new, surprising and of value. For a machine to be deemed truly creative requires one extra step: its contribution should be more than an expression of the coder's creativity or that of the person who built the data set. That is the challenge Ada Lovelace believed was insurmountable.

2

CREATING CREATIVITY

The chief enemy of creativity is good sense.
Pablo Picasso

The value placed on creativity in modern times has led to a range of writers and thinkers trying to articulate what it is, how to stimulate it, and why it is important. It was while sitting on a committee at the Royal Society assessing what impact machine learning was likely to have on society in the coming decades that I first encountered the theories of the cognitive scientist Margaret Boden. Her ideas on creativity struck me as the most relevant when it came to addressing or evaluating creativity in machines.

Boden is an original thinker who over the decades has managed to fuse many different disciplines: philosopher, psychologist, physician, AI expert and cognitive scientist. In her eighties now, with white hair flying like sparks and an ever-active brain, she is enjoying engaging enthusiastically with the prospect of what these 'tin cans', as she likes to call computers, might be capable of. To this end, she has identified three different types of human creativity.

Exploratory creativity involves taking what is already there and exploring its outer edges, extending the limits of what is possible while remaining bound by the rules. Bach's music is the culmination of a journey Baroque composers embarked on to explore tonality by weaving together different voices. His preludes and fugues push the boundaries of what is possible before breaking the genre open and entering the Classical era of Mozart and Beethoven. Renoir and Pissarro reconceived how we could visualise nature and the world around us, but it was Claude Monet who really pushed the boundaries, painting his water lilies over and over until his flecks of colour dissolved into a new form of abstraction.

Mathematics revels in this type of creativity. The classification of finite simple groups is a tour de force of exploratory creativity. Starting from the simple definition of a group of symmetries – a structure defined by four simple axioms – mathematicians spent 150 years producing a list of every conceivable element of symmetry, culminating in the discovery of the Monster Symmetry Group, which has more symmetries than there are atoms in the Earth and yet fits into no pattern of other groups. This form of mathematical creativity involves pushing the limits while adhering to the rules of the game. It is like the explorer who thrusts into the unknown but is still bound by the limits of our planet.

Boden believes that exploration accounts for 97 per cent of human creativity. This is the sort of creativity that computers excel at: pushing a pattern or set of rules to the extremes is perfect for a computational mechanism that can perform many more calculations than the human brain. But is it enough? When we think of truly original creative acts, we generally imagine something more utterly unexpected.

The second sort of creativity involves *combination*. Think of how an artist might take two completely different constructs and seek to combine them. Often the rules governing one world will suggest an interesting new framework for the other. Combination is a very powerful tool in the realm of mathematical creativity.

The eventual solution of the Poincaré Conjecture, which describes the possible shapes of our universe, was arrived at by applying very different tools to understand flow over surfaces. It was the creative genius of Grigori Perelman which realised that the way a liquid flows over a surface could unexpectedly help to classify the possible surfaces that might exist.

My own research takes tools from number theory to understand primes and applies them to classify possible symmetries. The symmetries of geometric objects at first sight don't look anything like numbers. But applying the language that has helped us to navigate the mysteries of the primes and replacing primes by symmetrical objects has revealed surprising new insights into the theory of symmetry.

The arts have also benefited greatly from this form of cross-fertilisation. Philip Glass took ideas he learned from working with Ravi Shankar and used them to create the additive process that is at the heart of his minimalist music. Zaha Hadid combined her knowledge of architecture with her love of the pure forms of the Russian painter Kasimir Malevich to create a unique style of curvaceous buildings. In cooking, too, creative master chefs have fused cuisines from opposite ends of the globe.

There are interesting hints that this sort of creativity might also be perfect for the world of AI. Take an algorithm that plays the blues and combine it with the music of Boulez and you will end up with a strange hybrid composition that might just create a new sound world. Of course, it could also be a dismal cacophony. The coder needs to find two genres that can be fused algorithmically in an interesting way.

It is Boden's third form of creativity that is the more mysterious and elusive, and that is *transformational* creativity. This describes those rare moments that are complete game changers. Every art form has these gear shifts. Think of Picasso and Cubism, Schoenberg and atonality, Joyce and modernism. They are like phase changes, when water suddenly goes from a liquid to a gas. This was the image Goethe hit on when he sought to describe wrestling for two years with how to write *The Sorrows of Young Werther*, only for a chance event to act as a sudden catalyst: 'At that instant, the plan of Werther was found; the whole shot together from all directions, and became a solid mass, as the water in a vase, which is just at the freezing point, is changed by the slightest concussion into ice.'

Quite often these transformational moments hinge on changing the rules of the game, or dropping an assumption that previous generations had been working under. The square of a number is always positive. All molecules come in long lines not chains. Music must be written inside a harmonic scale structure. Faces have eyes on either side of the nose. At first glance it would seem hard to program such a decisive break, and yet there is a meta-rule for this type of creativity. You start by dropping constraints and see what emerges. The art, the creative act, is to choose what to drop or what fresh constraint to introduce such that you end up with a new thing of value.

If I were asked to identify a transformational moment in mathematics, the creation of the square root of minus one in the mid-sixteenth century would be a good candidate. This was a number that many mathematicians believed did not exist. It was referred to as an imaginary number (a derogatory term Descartes came up with to indicate that of course there was no such thing). And yet its creation did not contradict previous mathematics. It turned out it had been our mistake to exclude it. How can a computer come up with the concept of the square root of minus one when the data it is fed will tell it that there is no number whose square can be negative? A truly creative act sometimes requires us to step outside the system and create a new reality. Can a complex algorithm do that?

The emergence of the Romantic movement in music is in many ways a catalogue of rule breaking. Instead of moving between close key signatures as Classical composers had done, new upstarts like Schubert chose to shift key in ways that deliberately broke expectations. Schumann left chords unresolved that Haydn or Mozart would have felt the need to complete. Chopin in turn composed dense moments of chromatic runs and challenged rhythmic expectations with his unusual accented passages and bending of tempos. The move from one musical movement to another: from Medieval to Baroque to Classical to Romantic to Impressionist to Expressionist and beyond is a story of breaking the rules. Each movement is dependent on the one before to appreciate its creativity. It almost goes without saying that historical context plays an important role in allowing us to define something as new. Creativity is not an absolute but a relative activity. We are creative within our culture and frame of reference.

Can a computer initiate this kind of phase change and move us into a new musical or mathematical state? That seems a challenge. Algorithms learn how to act based on the data they interact with. Won't this mean that they will always be condemned to producing more of the same?

As Picasso once said: 'The chief enemy of creativity is good sense.' That sounds on the face of it very much against the spirit of the machine. And yet you can program a system to behave irrationally. You can create a meta-rule that will instruct it to change course. As we shall see, this is in fact something machine learning is quite good at.

Can creativity be taught?

Many artists like to fuel their own creation myth, appealing to external forces as responsible for their creativity. In Ancient Greece poets were said to be possessed by the muses, who breathed inspiration into the minds of men, sometimes sending them insane in the process. For Plato 'a poet is holy, and never able to compose until he has become inspired, and is beside himself and reason is no longer in him . . . for no art does he utter but by power divine'. Ramanujan, the great Indian mathematician, likewise attributed his great insights to ideas he received in his dreams from his family goddess Namagiri. Is creativity a form of madness or a gift of the divine?

One of my mathematical heroes, Carl Friedrich Gauss, was one of the worst at covering his creative tracks. Gauss is credited with creating modern number theory with the publication in 1798 of one of the great mathematical works of all time: *Disquisitiones arithmeticae*. When people tried to read the book to uncover where he got his ideas, they were mystified. The work has been described as a book of seven seals. Gauss seems to pull ideas like rabbits out of a hat, without ever really giving us an inkling of how he achieved this magic. Later, when challenged, he retorted that an architect does not leave up the scaffolding after the house is complete. Gauss, like Ramanujan, attributed one revelation to 'the Grace of God', saying he was 'unable to name the nature of the thread which connected what I previously knew with that which made my success possible'.

Yet the fact that an artist may be unable to articulate where their ideas came from does not mean that they followed no rules. Art is a conscious expression of the myriad of logical gates that make up our unconscious thought processes. There was of course a thread of logic that connected Gauss's thoughts: it was just hard for him to articulate what he was up to – or perhaps he wanted to preserve the mystery, to fuel his image as a creative genius. Coleridge's claim that the drug-induced vision of Kubla Khan came to him in its entirety belies all the preparatory material that shows the poet working on the ideas before that fateful day when he was interrupted by the person from Porlock. Of course, this makes for a good story. Even my own account of creation will focus on the flash of inspiration rather than the years of preparatory work I put in.

We have an awful habit of romanticising creative genius. The solitary artist working in isolation is frankly a myth. In most instances what looks like a step change is actually a continuous growth. Brian Eno talks about the idea of 'scenius', not genius, to acknowledge the community out of which creative intelligence often emerges. The American writer Joyce Carol Oates agrees: 'Creative work, like scientific work, should be greeted as a communal effort – an attempt by an individual to give voice to many voices, an attempt to synthesize and explore and analyze.'

What does it take to stimulate creativity? Might it be possible to program it into a machine? Are there rules we can follow to become creative? Can creativity, in other words, be a learned skill? Some would say that to teach or program is to show people how to imitate what has gone before, and that imitation and rule following are both incompatible with creativity. And yet we have examples of creative individuals all around us who have studied and learned and improved their skills. If we study what they do, could we imitate them and ultimately become creative ourselves?

These are questions I find myself asking every new semester. To receive their PhDs, doctoral candidates in mathematics have to create a new mathematical construct. They have to come up with something that has never been done before. I am tasked with teaching them how to do that. Of course, at some level they have been training to do this to a certain extent already. Solving problems involves personal creativity even if the answer is already known.

That training is an absolute prerequisite for the jump into the unknown. By rehearsing how others have come to their breakthroughs you hope to provide the environment to foster your own creativity. And yet that jump is far from guaranteed. I can't take anyone off the street and teach them to be a creative mathematician. Maybe with ten years of training we could get there, but not every brain seems to be able to achieve mathematical creativity. Some people appear to be able to achieve creativity in one field but not another, yet it is difficult to understand what makes one brain a chess champion and another a Nobel Prize-winning novelist.

Margaret Boden recognises that creativity isn't just about being Shakespeare or Einstein. She distinguishes between what she calls 'psychological creativity' and 'historical creativity'. Many of us achieve acts of personal creativity that may be novel to us but historically old news. These are what Boden calls moments of psychological creativity. It is by repeated acts of personal creativity that ultimately one hopes to produce something that is recognised by others as new and of value. While historical creativity is rare, it emerges from encouraging psychological creativity.

My recipe for eliciting creativity in students follows the three modes of creativity Boden identified. Exploration is perhaps the most obvious path. First understand how we've come to the place we are now and then try to push the boundaries just a little bit further. This involves deep immersion in what we have created to date. Out of that deep understanding might emerge something never seen before. It is often important to impress on students that there isn't very often some big bang that resounds with the act of creation. It is gradual. As Van Gogh wrote: 'Great things are not done by impulse but by small things brought together.'

Boden's second strategy, combinational creativity, is a powerful weapon, I find, in stimulating new ideas. I often encourage students to attend seminars and read papers in subjects that don't appear to connect with the problem they are tackling. A line of thought from a disparate bit of the mathematical universe might resonate with the problem at hand and stimulate a new idea. Some of the most creative bits of science are happening today at the junctions between the disciplines. The more we can come out of our silos and share our ideas and problems, the more creative we are likely to be. This is where a lot of the low-hanging fruit is to be found.

At first sight transformational creativity seems hard to harness as a strategy. But again the goal is to test the status quo by dropping some of the constraints that have been put in place. Try seeing what happens if we change one of the basic rules we have accepted as part of the fabric of our subject. These are dangerous moments because you can collapse the system, but this brings me to one of the most important ingredients needed to foster creativity – and that is embracing failure.

Unless you are prepared to fail, you will not take the risks that will allow you to break out and create something new. This is why our education system and our business environment, both realms that abhor failure, are often terrible environments for fostering creativity. It is important to celebrate the failures as much as the successes in my students. Sure, the failures won't make it into the PhD thesis, but we learn so much from failure. When I meet with my students I repeat again and again Beckett's call to 'Fail, fail again, fail better.'

Are these strategies that can be written into code? In the past the top-down approach to coding meant there was little prospect of creativity in the output of the code. Coders were never too surprised by what their algorithms produced. There was no room for experimentation or failure. But this all changed recently: because an algorithm, built on code that learns from its failures, did something that was new, shocked its creators, and had incredible value. This algorithm won a game that many believed was beyond the abilities of a machine to master. It was a game that required creativity to play.

It was news of this breakthrough that triggered my recent existential crisis as a mathematician.

3

READY, STEADY, GO

We construct and construct,
but intuition is still a good thing.

Paul Klee

People often compare mathematics to playing chess. There certainly are connections, but when Deep Blue beat the best chessmaster the human race could offer in 1997, it did not lead to the closure of mathematics departments. Although chess is a good analogy for the formal quality of constructing a proof, there is another game that mathematicians have regarded as much closer to the creative and intuitive side of being a mathematician, and that is the Chinese game of Go.

I first discovered Go when I visited the mathematics department at Cambridge as an undergraduate to explore whether to do my PhD with the amazing group that had helped complete the classification of finite simple groups, a sort of Periodic Table of Symmetry. As I sat talking to John Conway and Simon Norton, two of the architects of this great project, about the future of mathematics, I kept being distracted by students at the next table furiously slamming black and white stones onto a large 19x19 grid carved into a wooden board.

Eventually I asked Conway what they were doing. 'That's Go. It's the oldest game that is still being played to this day.' In contrast to the war-like quality of chess, he explained, Go was a game of territory. Players take it in turn to place white and black pieces or stones onto the 19x19 grid. If you manage to surround a collection of your opponent's stones with your own, you capture your opponent's stones. The winner is the player who has captured the most stones by the end of the game. It sounded rather simple. The subtlety of the game, Conway explained, is that as you try to surround your opponent, you must avoid having your own stones captured.

'It's a bit like mathematics: simple rules that give rise to beautiful complexity.' It was while watching the game evolve between two experts as they drank coffee in the common room that Conway discovered that the endgame was behaving like a new sort of number that he christened 'surreal numbers'.

I've always been fascinated by games. Whenever I travel abroad I like to learn and bring back the game locals like to play. So when I got back from the wild outreaches of Cambridge to the safety of my home in Oxford I decided to buy Go from the local toy shop to see what it was that was obsessing these students. As I began to explore the game with one of my fellow students in Oxford, I realised how subtle it was. It was hard to identify a clear strategy that would help me win. And as more stones were laid down on the board, the game seemed to get more complicated, unlike chess, where as pieces are gradually removed the game starts to simplify.

The American Go Association estimates that it would take a number with 300 digits to count the number of games of Go that are legally possible. In chess the computer scientist Claude Shannon estimated that a number with 120 digits (now called the Shannon number) would suffice. These are not small numbers in either case, but they give you a sense of the wide range of possible permutations.

I had played a lot of chess as a kid. I enjoyed working through the logical consequences of a proposed move. It appealed to the mathematician that was growing inside me. The tree of possibilities in chess branches in a controlled manner, making it manageable for a computer and even a human to analyse the implications of going down different branches. In contrast Go just doesn't seem like a game that would allow you to work out the logical implications of a future move. Navigating the tree of possibilities quickly becomes impossible. That's not to say that a Go player doesn't follow through the logical consequences of their next move, but this seems to be combined with a more intuitive feel for the pattern of play.

The human brain is acutely attuned to finding structure and pattern if there is one in a visual image. A Go player can look at the lie of the stones and tap into the brain's ability to pick out these patterns and exploit them in planning the next move. Computers have traditionally always struggled with vision. It is one of the big hurdles that engineers have wrestled with for decades.

The human brain's highly developed sense of visual structure has been honed over millions of years and has been key to our survival. Any animal's ability to survive depends in part on its ability to pick out structure in the visual mess that Nature confronts us with. A pattern in the chaos of the jungle is likely to be evidence of the presence of another animal – and you'd better take notice cos that animal might eat you (or maybe you could eat it). The human code is extremely good at reading patterns, interpreting how they might develop, and responding appropriately. It is one of our key assets, and it plays into our appreciation for the patterns in music and art.

It turns out that pattern recognition is precisely what I do as a mathematician when I venture into the unexplored reaches of the mathematical jungle. I can't rely on a simple step-by-step logical analysis of the local environment. That won't get me very far. It has to be combined with an intuitive feel for what might be out there. That intuition is built up by time spent exploring the known space. But it is often hard to articulate logically why you might believe that there is interesting territory out there to explore. A conjecture in mathematics is by its nature not yet proved, but the mathematician who has made the conjecture has built up a feeling that the mathematical statement they have made may have some truth to it. Observation and intuition go hand in hand as we navigate the thickets and seek to carve out a new path.

A mathematician who can make a good conjecture will often garner more respect than one who joins up the logical dots to reveal the truth of the conjecture. In the game of Go the final winning position is in some respects the conjecture and the plays are the logical moves on your way to proving that conjecture. But it is devilishly hard to spot the patterns along the way.

And so, although chess has been useful to help explain some aspects of mathematics, the game of Go has always been held up as far closer in spirit to the way mathematicians actually go about their business. That's why mathematicians weren't too worried when Deep Blue beat the best humans could offer at chess. The real challenge was the game of Go. For decades people have been claiming that the game of Go can never be played by a computer. Like all good absolutes, it invited creative coders to test that proposition. But even a junior player appeared to be able to outplay even the most complex algorithms. And so mathematicians happily hid behind the cover that Go was providing them. If a computer couldn't play Go then there was no chance it could play the even subtler and more ancient game of mathematics.

But just as the Great Wall of China was eventually breached, my defensive wall has just crumbled in spectacular fashion.

Game Boy extraordinaire

At the beginning of 2016 it was announced that a program had been created to play Go that its developers were confident could hold its own against the best humans had to offer. Go players around the world were extremely sceptical, given the failure of past efforts. So the company that developed the program offered a challenge. It set up a public contest with a huge prize and invited one of the world's leading Go players to take up the challenge. An international champion, Lee Sedol from Korea, stepped forward. The competition would be played over five games with the winner taking home a prize of one million dollars. The name of Sedol's challenger: AlphaGo.

AlphaGo is the brainchild of Demis Hassabis. Hassabis was born in London in 1976 to a Greek Cypriot father and a mother from Singapore. Both parents are teachers and what Hassabis describes as bohemian technophobes. His sister and brother went the creative route, one becoming a composer, the other choosing creative writing. So Hassabis isn't quite sure where his geeky scientific side came from. But as a kid Hassabis was someone who quickly marked himself out as gifted and talented, especially when it came to playing games. His abilities at chess were such that at eleven he was the second-highest-ranked child of his age in the world.

But then at an international match in Liechtenstein that year Hassabis had an epiphany: what on earth were they all doing? The hall was full of so many great minds exploring the logical intricacies of this great game. And yet Hassabis suddenly recognised the total futility of such a project. In a radio interview on the BBC he admitted thinking at the time: 'We were wasting our minds. What if we used that brain power for something more useful like solving cancer?'

His parents were pretty shocked when after the tournament (which he narrowly lost after battling for ten hours with the adult Dutch world champion) he announced that he was giving up chess competitions. Everyone had thought this was going to be his life. But those years playing chess weren't wasted. A few years earlier he'd used the £200 prize money he'd won for beating a US opponent, Alex Chang, to buy his first computer: a ZX Spectrum. That computer sparked his obsession with getting machines to do the thinking for him.

Hassabis soon graduated on to a Commodore Amiga, which could be programmed to play the games he enjoyed. Chess was still too complicated, but he managed to program the Commodore to play Othello, a game that looks rather similar to Go with black and white stones that get flipped when they are trapped between stones of the opposite colour. It's not a game that merits grandmasters, so he tried his program out on his younger brother. It beat him every time.

This was classic 'if . . . , then . . .' programming: he needed to code in by hand the response to each of his opponent's moves. It was: 'If your opponent plays that move, then reply with this move.' The creativity all came from Hassabis and his ability to see what the right responses were to win the game. It still felt a bit like magic though. Code up the right spell and then, rather like the Sorcerer's Apprentice, the Commodore would go through the work of winning the game.

Hassabis raced through school, culminating with an offer from Cambridge to study computer science at the age of sixteen. He'd set his heart on Cambridge after seeing Jeff Goldblum in the film *The Race for the Double Helix*. 'I thought, is this what goes on at Cambridge? You go there and you invent DNA in the pub? Wow.'

Cambridge wouldn't let him start his degree at the age of sixteen, so he had to defer for a year. To fill his time he won a place working for a game developer after having come second in a competition run by *Amiga Power* magazine. While he was there, he created his own game, Theme Park, where players had to build and run their own theme park. The game was hugely successful, selling several million copies and winning a Golden Joystick award. With enough funds to finance his time at university, Hassabis set off for Cambridge.

His course introduced him to the greats of the AI revolution: Alan Turing and his test for intelligence, Arthur Samuel and his program to play draughts, John McCarthy, who coined the term artificial intelligence, Frank Rosenblatt and his first experiments with neural networks. These were the shoulders on which Hassabis aspired to stand. It was while sitting in his lectures at Cambridge that he heard his professor repeating the mantra that a computer could never play Go because of the game's creative and intuitive characteristics. This was like a red rag to the young Hassabis. He left Cambridge determined to prove his professor wrong.

His idea was that rather than trying to write a program himself that could play Go, he would write a meta-program that would be responsible for writing the program that would play Go. It sounded a crazy idea, but the point was that the meta-program would be created so that as the Go-playing program played more and more games it would learn from its mistakes.

Hassabis had learned about a similar idea implemented by the artificial-intelligence researcher Donald Michie in the 1960s. Michie had written an algorithm called 'MENACE' that learned from scratch the best strategy to play noughts and crosses. (MENACE stood for Machine Educable Noughts And Crosses Engine.) To demonstrate the algorithm, Michie had rigged up 304 matchboxes representing all the possible layouts of noughts and crosses encountered while playing. Each matchbox was filled with different-coloured balls to represent possible moves. Balls were removed or added to the boxes to punish losses or reward wins. As the algorithm played more and more games, the reassignment of the balls eventually led to an almost perfect strategy for playing. It was this idea of learning from your mistakes that Hassabis wanted to use to train an algorithm to play Go.

Hassabis had a good model to base his strategy on. A newborn baby does not have a brain that is pre-programmed to cope with making its way through life. It is programmed instead to learn as it interacts with its environment.

If Hassabis was going to tap into the way the brain learned to solve problems, then knowing how the brain works was clearly going to help in his dream of creating a program to play Go. So he decided to do a PhD in neuroscience at University College London. It was during coffee breaks from lab work that Hassabis started discussing with a neuroscientist, Shane Legg, his plans to create a company to try out his ideas. It shows the low status of AI even a decade ago that they never admitted to their professors their dream to dedicate their lives to AI. But they felt they were on to something big, so in September 2010 the two scientists decided to create a company with Mustafa Suleyman, a friend of Hassabis from childhood. DeepMind was incorporated.

The company needed money but initially Hassabis just couldn't raise any capital. Pitching on a platform that they were going to play games and solve intelligence did not sound serious to most investors. A few, however, did see the vision. Among those who put money in right at the outset were Elon Musk and Peter Thiel. Thiel had never invested outside Silicon Valley and tried to persuade Hassabis to relocate to the West Coast. A born-and-bred Londoner, Hassabis held his ground, insisting that there was more untapped talent in London that could be exploited. Hassabis remembers a crazy conversation he had with Thiel's lawyer. 'Does London have law on IP?' she asked innocently. 'I think they thought we were coming from Timbuctoo!' The founders had to give up a huge amount of stock to the investors, but they had their money to start trying to crack AI.

The challenge of creating a machine that could learn to play Go still felt like a distant dream. They set their sights at first on a seemingly less cerebral goal: playing 1980s Atari games. Atari is probably responsible for a lot of students flunking courses in the late 1970s and early 1980s. I certainly remember wasting a huge amount of time playing the likes of Pong, Space Invaders and Asteroids on a friend's Atari 2600 console. The console was one of the first whose hardware could play multiple games that were loaded via a cartridge. It allowed a whole range of different games to be developed over time. Previous consoles could only play games that had been physically programmed into the units.

One of my favourite Atari games was called Breakout. A wall of coloured bricks was at the top of the screen and you controlled a paddle at the bottom that could be moved left or right using a joystick. A ball would bounce off the paddle and head towards the bricks. Each time it hit a brick, the brick would disappear. The aim was to clear the bricks. The yellow bricks at the bottom of the wall scored one point. The red bricks on top got you seven points. As you cleared blocks, the paddle would shrink and the ball would speed up to make the game play harder.

We were particularly pleased one afternoon when we found a clever way to hack the game. If you dug a tunnel up through the bricks on the edge of the screen, once the ball made it through to the top it bounced back and forward off the top of the screen and the upper high-scoring bricks, gradually clearing the wall. You could sit back and watch until the ball eventually came back down through the wall. You just had to be ready with the paddle to bat the ball back up again. It was a very satisfying strategy!

Hassabis and the team he was assembling also spent a lot of time playing computer games in their youth. Their parents may be happy to know that the time and effort they put into those games did not go to waste. It turned out that Breakout was a perfect test case to see if the team at DeepMind could program a computer to learn how to play games. It would have been a relatively straightforward job to write a program for each individual game. Hassabis and his team were going to set themselves a much greater challenge.

They wanted to write a program that would receive as an input the state of the pixels on the screen and the current score and set it to play with the goal of maximising the score. The program was not told the rules of the game: it had to experiment randomly with different ways of moving the paddle in Breakout or firing the laser cannon at the descending aliens of Space Invaders. Each time it made a move it could assess whether the move had helped increase the score or had had no effect.

The code implements an idea dating from the 1990s called reinforcement learning, which aims to update the probability of actions based on the effect on a reward function or score. For example, in Breakout the only decision is whether to move the paddle at the bottom left or right. Initially the choice will be 50:50. But if moving the paddle randomly results in it hitting the ball, then a short time later the score goes up. The code then recalibrates the probability of whether to go left or right based on this new information. It will increase the chance of heading in the direction towards which the ball is heading. The new feature was to combine this learning with neural networks that would assess the state of the pixels to decide what features were correlating to the increase in score.

At the outset, because the computer was just trying random moves, it was terrible, hardly scoring anything. But each time it made a random move that bumped up the score, it would remember the move and reinforce the use of such a move in future. Gradually the random moves disappeared and a more informed set of moves began to emerge, moves that the program had learned through experiment seemed to boost its score.

It's worth watching the supplementary video the DeepMind team included in the paper they eventually wrote. It shows the program learning to play Breakout. At first you see it randomly moving the paddle back and forward to see what will happen. Then, when the ball finally hits the paddle and bounces back and hits a brick and the score goes up, the program starts to rewrite itself. If the pixels of the ball and the pixels of the paddle connect that seems to be a good thing. After 400 game plays it's doing really well, getting the paddle to continually bat the ball back and forward.

The shock for me came when you see what it discovered after 600 games. It found our hack! I'm not sure how many games it took us as kids to find this trick, but judging by the amount of time I wasted with my friend it could well have been more. But there it is. The program manipulated the paddle to tunnel its way up the sides, such that the ball would be stuck in the gap between the top of the wall and the top of the screen. At this point the score goes up very fast without the computer's having to do very much. I remember my friend and I high-fiving when we'd discovered this trick. The machine felt nothing.

By 2014, four years after the creation of DeepMind, the program had learned how to outperform humans on twenty-nine of the forty-nine Atari games it had been exposed to. The paper the team submitted to *Nature* detailing their achievement was published in early 2015. To be published in *Nature* is one of the highlights of a scientist's career. But their paper achieved the even greater accolade of being featured as the cover story of the whole issue. The journal recognised that this was a huge moment for artificial intelligence.

It has to be reiterated what an amazing feat of programming this was. From just the raw data of the state of the pixels and the changing score, the program had changed itself from randomly moving the paddle of Breakout back and forth to learning that tunnelling the sides of the wall would win you the top score. But Atari games are hardly on a par with the ancient game of Go. Hassabis and his team at DeepMind decided they were ready to create a new program that could take it on.

It was at this moment that Hassabis decided to sell the company to Google. 'We weren't planning to, but three years in, focused on fundraising, I had only ten per cent of my time for research,' he explained in an interview in *Wired* at the time. 'I realised that there's maybe not enough time in one lifetime to both build a Google-sized company and solve AI. Would I be happier looking back on building a multi-billion business or helping solve intelligence? It was an easy choice.' The sale put Google's firepower at his fingertips and provided the space for him to create code to realise his goal of solving Go . . . and then intelligence.

First blood

Previous computer programs built to play Go had not come close to playing competitively against even a pretty good amateur, so most pundits were highly sceptical of DeepMind's dream to create code that could get anywhere near an international champion of the game. Most people still agreed with the view expressed in *The New York Times* by the astrophysicist Piet Hut after DeepBlue's success at chess in 1997: 'It may be a hundred years before a computer beats humans at Go – maybe even longer. If a reasonably intelligent person learned to play Go, in a few months he could beat all existing computer programs. You don't have to be a Kasparov.'

Just two decades into that hundred years, the DeepMind team believed they might have cracked the code. Their strategy of getting algorithms to learn and adapt appeared to be working, but they were unsure quite how powerful the emerging algorithm really was. So in October 2015 they decided to test-run their program in a secret competition against the current European champion, the Chinese-born Fan Hui.

AlphaGo destroyed Fan Hui five games to nil. But the gulf between European players of the game and those in the Far East is huge. The top European players, when put in a global league, rank in the 600s. So, although it was still an impressive achievement, it was like building a driverless car that could beat a human driving a Ford Fiesta round Silverstone then trying to challenge Lewis Hamilton in a Grand Prix.

Certainly when the press in the Far East heard about Fan Hui's defeat they were merciless in their dismissal of how meaningless the win was for AlphaGo. Indeed, when Fan Hui's wife contacted him in London after the news got out, she begged her husband not to go online. Needless to say he couldn't resist. It was not a pleasant experience to read how dismissive the commentators in his home country were of his credentials to challenge AlphaGo.

Fan Hui credits his matches with AlphaGo with teaching him new insights into how to play the game. In the following months his ranking went from 633 to the 300s. But it wasn't only Fan Hui who was learning. Every game AlphaGo plays affects its code and changes it to improve its play next time around.

It was at this point that the DeepMind team felt confident enough to offer their challenge to Lee Sedol, South Korea's eighteen-time world champion and a formidable player of the game.

The match was to be played over five games scheduled between 9 and 15 March 2016 at the Four Seasons hotel in Seoul, and would be broadcast live across the internet. The winner would receive a prize of a million dollars. Although the venue was public, the precise location within the hotel was kept secret and was isolated from noise – not that AlphaGo was going to be disturbed by the chitchat of the press and the whispers of curious bystanders. It would assume a perfect Zen-like state of concentration wherever it was placed.

Sedol wasn't fazed by the news that he was up against a machine that had beaten Fan Hui. Following Fan Hui's loss he had declared: 'Based on its level seen . . . I think I will win the game by a near landslide.'

Although he was aware of the fact that the machine he would be playing was learning and evolving, this did not concern him. But as the match approached, you could hear doubts beginning to creep into his view of whether AI will ultimately be too power-ful for humans to defeat it even in the game of Go. In February he stated: 'I have heard that DeepMind's AI is surprisingly strong and getting stronger, but I am confident that I can win . . . at least this time.'

Most people still felt that despite great inroads into pro-gramming, an AI Go champion was still a distant goal. Rémi Coulom, the creator of Crazy Stone, the only program to get close to playing Go at any high standard, was still predicting another decade before computers would beat the best humans at the game.

As the date for the match approached, the team at DeepMind felt they needed someone to really stretch AlphaGo and to test it for any weaknesses. So they invited Fan Hui back to play the machine going into the last few weeks. Despite having suffered a 5–0 defeat and being humiliated by the press back in China, he was keen to help out. Perhaps a bit of him felt that if he could help make AlphaGo good enough to beat Sedol, it would make his defeat less humiliating.

As Fan Hui played he could see that AlphaGo was extremely strong in some areas but he managed to reveal a weakness that the team was not aware of. There were certain configurations in which it seemed to completely fail to assess who had con-trol of the game, often becoming totally delusional that it was winning when the opposite was true. If Sedol tapped into this weakness, AlphaGo wouldn't just lose, it would appear extremely stupid.

The DeepMind team worked around the clock trying to fix this blind spot. Eventually they just had to lock down the code as it was. It was time to ship the laptop they were using to Seoul.

The stage was set for a fascinating duel as the players, or at least one player, sat down on 9 March to play the first of the five games.

'Beautiful. Beautiful. Beautiful'

It was with a sense of existential anxiety that I fired up the You-Tube channel broadcasting the matches that Sedol would play against AlphaGo and joined 280 million other viewers to see humanity take on the machines. Having for years compared creating mathematics to playing the game of Go, I had a lot on the line.

Lee Sedol picked up a black stone and placed it on the board and then waited for the response. Aja Huang, a member of the DeepMind team, would play the physical moves for AlphaGo. This, after all, was not a test of robotics but of artificial intelligence. Huang stared at AlphaGo's screen, waiting for its response to Sedol's first stone. But nothing came.

We all stared at our screens wondering if the program had crashed! The DeepMind team was also beginning to wonder what was up. The opening moves are generally something of a formality. No human would think so long over move 2. After all, there was nothing really to go on yet. What was happening? And then a white stone appeared on the computer screen. It had made its move. The DeepMind team breathed a huge sigh of relief. We were off! Over the next couple of hours the stones began to build up across the board.

One of the problems I had as I watched the game was assessing who was winning at any given point in the game. It turns out that this isn't just because I'm not a very experienced Go player. It is a characteristic of the game. Indeed, this is one of the main reasons why programming a computer to play Go is so hard. There isn't an easy way to turn the current state of the game into a robust scoring system of who leads by how much.

Chess, by contrast, is much easier to score as you play. Each piece has a different numerical value which gives you a simple first approximation of who is winning. Chess is destructive. One by one pieces are removed so the state of the board simplifies as the game proceeds. But Go increases in complexity as you play. It is constructive. The commentators kept up a steady stream of observations but struggled to say if anyone was in the lead right up until the final moments of the game.

What they were able to pick up quite quickly was Sedol's opening strategy. If AlphaGo had learned to play on games that had been played in the past, then Sedol was working on the principle that it would put him at an advantage if he disrupted the expectations it had built up by playing moves that were not in the conventional repertoire. The trouble was that this required Sedol to play an unconventional game – one that was not his own.

It was a good idea but it didn't work. Any conventional machine programmed on a database of accepted openings wouldn't have known how to respond and would most likely have made a move that would have serious consequences in the grand arc of the game. But AlphaGo was not a conventional machine. It could assess the new moves and determine a good response based on what it had learned over the course of its many games. As David Silver, the lead programmer on AlphaGo, explained in the lead-up to the match: 'AlphaGo learned to discover new strategies for itself, by playing millions of games between its neural networks, against themselves, and gradually improving.' If anything, Sedol had put himself at a disadvantage by playing a game that was not his own.

As I watched I couldn't help feeling for Sedol. You could see his confidence draining out of him as it gradually dawned on him that he was losing. He kept looking over at Huang, the DeepMind representative who was playing AlphaGo's moves, but there was nothing he could glean from Huang's face. By move 186 Sedol had to recognise that there was no way to overturn the advantage AlphaGo had built up on the board. He placed a stone on the side of the board to indicate his resignation.

By the end of day one it was: AlphaGo 1 Humans 0. Sedol

admitted at the press conference that day: 'I was very surprised because I didn't think I would lose.'

But it was game 2 that was going to truly shock not just Sedol but every human player of the game of Go. The first game was one that experts could follow and appreciate why AlphaGo was playing the moves it was. They were moves a human champion would play. But as I watched game 2 on my laptop at home, something rather strange happened. Sedol played move 36 and then retired to the roof of the hotel for a cigarette break. While he was away, AlphaGo on move 37 instructed Huang, its human representative, to place a black stone on the line five steps in from the edge of the board. Everyone was shocked.

The conventional wisdom is that during the early part of the game you play stones on the outer four lines. The third line builds up short-term territory strength on the edge of the board while playing on the fourth line contributes to your strength later in the game as you move into the centre of the board. Players have always found that there is a fine balance between playing on the third and fourth lines. Playing on the fifth line has always been regarded as suboptimal, giving your opponent the chance to build up territory that has both short- and long-term influence.

AlphaGo had broken this orthodoxy built up over centuries of competing. Some commentators declared it a clear mistake. Others were more cautious. Everyone was intrigued to see what Sedol would make of the move when he returned from his cigarette break. As he sat down, you could see him physically flinch as he took in the new stone on the board. He was certainly as shocked as all of the rest of us by the move. He sat there thinking for over twelve minutes. Like chess, the game was being played under time constraints. Using twelve minutes of your time was very costly. It is a mark of how surprising this move was that it took Sedol so long to respond. He could not understand what AlphaGo was doing. Why had the program abandoned the region of stones they were competing over?

Was this a mistake by AlphaGo? Or did it see something deep inside the game that humans were missing? Fan Hui, who had been given the role of one of the referees, looked down on the board. His initial reaction matched everyone else's: shock. And then he began to realise: 'It's not a human move. I've never seen a human play this move,' he said. 'So beautiful. Beautiful. Beautiful. Beautiful.'

Beautiful and deadly it turned out to be. Not a mistake but an extraordinarily insightful move. Some fifty moves later, as the black and white stones fought over territory from the lower left-hand corner of the board, they found themselves creeping towards the black stone of move 37. It was joining up with this stone that gave AlphaGo the edge, allowing it to clock up its second win. AlphaGo 2 Humans 0.

Sedol's mood in the press conference that followed was notably different. 'Yesterday I was surprised. But today I am speechless . . . I am in shock. I can admit that . . . the third game is not going to be easy for me.' The match was being played over five games. This was the game that Sedol needed to win to be able to stop AlphaGo claiming the match.

The human fight-back

Sedol had a day off to recover. The third game would be played on Saturday, 12 March. He needed the rest, unlike the machine. The first game had been over three hours of intense concentration. The second lasted over four hours. You could see the emotional toll that losing two games in a row was having on him.

Rather than resting, though, Sedol stayed up till 6 a.m. the next morning analysing the games he'd lost so far with a group of fellow professional Go players. Did AlphaGo have a weakness they could exploit? The machine wasn't the only one who could learn and evolve. Sedol felt he might learn something from his losses.

Sedol played a very strong opening to game 3, forcing AlphaGo to manage a weak group of stones within his sphere of influence on the board. Commentators began to get excited. Some said Sedol had found AlphaGo's weakness. But then, as one commentator posted: 'Things began to get scary. As I watched the game unfold and the realisation of what was happening dawned on me, I felt physically unwell.'

Sedol pushed AlphaGo to its limits but in so doing he revealed the hidden powers that the program seemed to possess. As the game proceeded, it started to make what commentators called lazy moves. It had analysed its position and was so confident in its win that it chose safe moves. It didn't care if it won by half a point. All that mattered was that it won. To play such lazy moves was almost an affront to Sedol, but AlphaGo was not programmed with any vindictive qualities. Its sole goal was to win the game. Sedol pushed this way and that, determined not to give in too quickly. Perhaps one of these lazy moves was a mistake that he could exploit.

By move 176 Sedol eventually caved in and resigned. AlphaGo 3 Humans 0. AlphaGo had won the match. Backstage, the Deep-Mind team was going through a strange range of emotions. They'd won the match, but seeing the devastating effect it was having on Sedol made it hard for them to rejoice. The million-dollar prize was theirs. They'd already decided to donate the prize, if they won, to a range of charities dedicated to promoting Go and science subjects as well as to Unicef. Yet their human code was causing them to empathise with Sedol's pain.

AlphaGo did not demonstrate any emotional response to its win. No little surge of electrical current. No code spat out with a resounding 'YES!' It is this lack of response that gives humanity hope and is also scary at the same time. Hope because it is this emotional response that is the drive to be creative and venture into the unknown: it was humans, after all, who'd programmed AlphaGo with the goal of winning. Scary because the machine won't care if the goal turns out to be not quite what its programmers had intended.

Sedol was devastated. He came out in the press conference and apologised:

> I don't know how to start or what to say today, but I think I would have to express my apologies first. I should have shown a better result, a better outcome, and better content in terms of the game played, and I do apologize for not being able to satisfy a lot of people's expectations. I kind of felt powerless.

But he urged people to keep watching the final two games. His goal now was to try to at least get one back for humanity.

Having lost the match, Sedol started game 4 playing far more freely. It was as if the heavy burden of expectation had been lifted, allowing him to enjoy his game. In sharp contrast to the careful, almost cautious play of game 3, he launched into a much more extreme strategy called 'amashi'. One commentator compared it to a city investor who, rather than squirrelling away small gains that accumulate over time, bet the whole bank.

Sedol and his team had stayed up all of Saturday night trying to reverse-engineer from AlphaGo's games how it played. It seemed to work on a principle of playing moves that incrementally increase its probability of winning rather than betting on the potential outcome of a complicated single move. Sedol had witnessed this when AlphaGo preferred lazy moves to win game 3. The strategy they'd come up with was to disrupt this sensible play by playing the risky single moves. An all-or-nothing strategy might make it harder for AlphaGo to score so easily.

AlphaGo seemed unfazed by this line of attack. Seventy moves into the game, commentators were already beginning to see that

AlphaGo had once again gained the upper hand. This was confirmed by a set of conservative moves that were AlphaGo's signal that it had the lead. Sedol had to come up with something special if he was going to regain the momentum.

If move 37 of game 2 was AlphaGo's moment of creative genius, move 78 of game 4 was Sedol's retort. He'd sat there for thirty minutes staring at the board, staring at defeat, when he suddenly placed a white stone in an unusual position, between two of AlphaGo's black stones. Michael Redmond, who was commentating on the YouTube channel, spoke for everyone: 'It took me by surprise. I'm sure that it would take most opponents by surprise. I think it took AlphaGo by surprise.'

It certainly seemed to. AlphaGo appeared to completely ignore the play, responding with a strange move. Within several more moves AlphaGo could see that it was losing. The DeepMind team stared at their screens behind the scenes and watched their creation imploding. It was as if move 78 short-circuited the program. It seemed to cause AlphaGo to go into meltdown as it made a whole sequence of destructive moves. This apparently is another characteristic of the way Go algorithms are programmed. Once they see that they are losing they go rather crazy.

Silver, the chief programmer, winced as he saw the next move AlphaGo was suggesting: 'I think they're going to laugh.' Sure enough, the Korean commentators collapsed into fits of giggles at the moves AlphaGo was now making. Its moves were failing the Turing Test. No human with a shred of strategic sense would make them. The game dragged on for a total of 180 moves, at which point AlphaGo put up a message on the screen that it had resigned. The press room erupted with spontaneous applause.

The human race had got one back. AlphaGo 3 Humans 1. The smile on Lee Sedol's face at the press conference that evening said it all. 'This win is so valuable that I wouldn't exchange it for anything in the world.' The press cheered wildly. 'It's because of the cheers and the encouragement that you all have shown me.'

Gu Li, who was commentating on the game in China, declared Sedol's move 78 as the 'hand of god'. It was a move that broke the conventional way to play the game and that was ultimately the key to its shocking impact. Yet this is characteristic of true human creativity. It is a good example of Boden's transformational creativity, whereby breaking out of the system you can find new insights.

At the press conference, Hassabis and Silver could not explain why AlphaGo had lost. They would need to go back and analyse why it had made such a lousy move in response to Sedol's move 78. It turned out that AlphaGo's experience in playing humans had led it to totally dismiss such a move as something not worth thinking about. It had assessed that this was a move that had only a one in 10,000 chance of being played. It seems as if it just had not bothered to learn a response to such a move because it had prioritised other moves as more likely and therefore more worthy of response.

Perhaps Sedol just needed to get to know his opponent. Perhaps over a longer match he would have turned the tables on AlphaGo. Could he maintain the momentum into the fifth and final game? Losing 3–2 would be very different from 4–1. The last game was still worth competing in. If he could win a second game, then it would sow seeds of doubt about whether AlphaGo could sustain its superiority.

But AlphaGo had learned something valuable from its loss. You play Sedol's one in 10,000 move now against the algorithm and you won't get away with it. That's the power of this sort of algorithm. It learns from its mistakes.

That's not to say it can't make new mistakes. As game 5 proceeded, there was a moment quite early on when AlphaGo seemed to completely miss a standard set of moves in response to a particular configuration that was building. As Hassabis tweeted from backstage: '#AlphaGo made a bad mistake early in the game (it didn't know a known tesuji) but now it is trying hard to claw it back . . . nail-biting.'

Sedol was in the lead at this stage. It was game on. Gradually AlphaGo did claw back. But right up to the end the DeepMind team was not exactly sure whether it was winning. Finally, on move 281 – after five hours of play – Sedol resigned. This time there were cheers backstage. Hassabis punched the air. Hugs and high fives were shared across the team. The win that Sedol had pulled off in game 4 had suddenly re-engaged their competitive spirit. It was important for them not to lose this last game.

Looking back at the match, many recognise what an extra-ordinary moment this was. Some immediately commented on its being an inflexion point for AI. Sure, all this machine could do was play a board game, and yet, for those looking on, its capability to learn and adapt was something quite new. Hassabis's tweet after winning the first game summed up the achievement: '#AlphaGo WINS!!!! We landed it on the moon.' It was a good comparison. Landing on the moon did not yield extraordinary new insights about the universe, but the technology that we developed to achieve such a feat has. Following the last game, AlphaGo was awarded an honorary professional 9 dan rank by the South Korean Go Association, the highest accolade for a Go player.

From hilltop to mountain peak

Move 37 of game 2 was a truly creative act. It was novel, certainly, it caused surprise, and as the game evolved it proved its value. This was exploratory creativity, pushing the limits of the game to the extreme.

One of the important points about the game of Go is that there is an objective way to test whether a novel move has value. Anyone can come up with a new move that appears creative. The art and challenge are in making a novel move that has some sort of value. How should we assess value? It can be very subjective and time-dependent. Something that is panned critically at the time of its release can be recognised generations later as a transformative creative act. Nineteenth-century audiences didn't know what to make of Beethoven's Symphony no. 5, and yet it is central repertoire now. During his lifetime Van Gogh could barely sell his paintings, trading them for food or painting materials, but now they go for millions. In Go there is a more tangible and immediate test of value: does it help you win the game? Move 37 won AlphaGo game 2. There was an objective measure that we could use to value the novelty of this move.

AlphaGo had taught the world a new way to play an ancient game. Analysis since the match has resulted in new tactics. The fifth line is now played early on, as we have come to understand that it can have big implications for the endgame. AlphaGo has gone on to discover still more innovative strategies. DeepMind revealed at the beginning of 2017 that its latest iteration had played online anonymously against a range of top-ranking professionals under two pseudonyms: Master and Magister. Human players were unaware that they were playing a machine. Over a few weeks it had played a total of sixty complete games. It won all sixty games.

But it was the analysis of the games that was truly insightful. Those games are now regarded as a treasure trove of new ideas. In several games AlphaGo played moves that beginners would have their wrists slapped for by their Go master. Traditionally you do not play a stone in the intersection of the third column and third row. And yet AlphaGo showed how to use such a move to your advantage.

Hassabis describes how the game of Go had got stuck on what mathematicians like to call a local maximum. If you look at the landscape I've drawn on page 42 then you might be at the top of peak A. From this height there is nowhere higher to go. This is

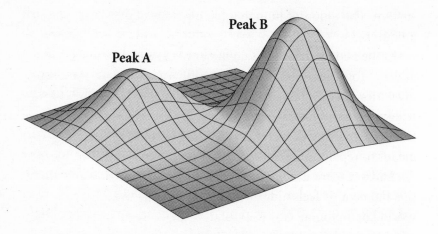

called a local maximum. If there were fog all around you, you'd think you were at the highest point in the land. But across the valley is a higher peak. To know this, you need the fog to clear. You need to descend from your peak, cross the valley and climb the higher peak.

The trouble with modern Go is that conventions had built up about ways to play that had ensured players hit peak A. But by breaking those conventions AlphaGo had cleared the fog and revealed an even higher peak B. It's even possible to measure the difference. In Go, a player using the conventions of peak A will in general lose by two stones to the player using the new strategies discovered by AlphaGo.

This rewriting of the conventions of how to play Go has happened at a number of previous points in history. The most recent was the innovative game play introduced by the legendary Go Seigen in the 1930s. His experimentation with ways of playing the opening moves revolutionised the way the game is played. But Go players now recognise that AlphaGo might well have launched an even greater revolution.

The Chinese Go champion Ke Jie recognises that we are in a new era: 'Humanity has played Go for thousands of years, and yet, as AI has shown us, we have not yet even scratched the surface. The union of human and computer players will usher in a new era.'

Ke Jie's compatriot Gu Li, winner of the most Go world titles, added: 'Together, humans and AI will soon uncover the deeper mysteries of Go.' Hassabis compares the algorithm to the Hubble telescope. This illustrates the way many view this new AI. It is a tool for exploring deeper, further, wider than ever before. It is not meant to replace human creativity but to augment it.

And yet there is something that I find quite depressing about this moment. It feels almost pointless to want to aspire to be the world champion at Go when you know there is a machine that you will never be able to beat. Professional Go players have tried to put a brave face on it, talking about the extra creativity that it has unleashed in their own play, but there is something quite soul-destroying about knowing that we are now second best to the machine. Sure, the machine was programmed by humans, but that doesn't really seem to make it feel better.

AlphaGo has since retired from competitive play. The Go team at DeepMind has been disbanded. Hassabis proved his Cambridge lecturer wrong. DeepMind has now set its sights on other goals: health care, climate change, energy efficiency, speech recognition and generation, computer vision. It's all getting very serious.

Given that Go was always my shield against computers doing mathematics, was my own subject next in DeepMind's cross hairs? To truly judge the potential of this new AI we are going to need to look more closely at how it works and dig around inside. The crazy thing is that the tools DeepMind is using to create the programs that might put me out of a job are precisely the ones that mathematicians have created over the centuries. Is this mathematical Frankenstein's monster about to turn on its creator?

4

ALGORITHMS, THE SECRET TO MODERN LIFE

The Analytical Engine weaves algebraic patterns,
just as the Jacquard loom weaves flowers and leaves.

Ada Lovelace

Our lives are completely run by algorithms. Every time we search for something on the internet, plan a journey with our GPS, choose a movie recommended by Netflix or pick a date online, we are being guided by an algorithm. Algorithms are steering us through the digital age, yet few people realise that they predate the computer by thousands of years and go to the heart of what mathematics is all about.

The birth of mathematics in Ancient Greece coincides with the development of one of the very first algorithms. In Euclid's *Elements*, alongside the proof that there are infinitely many prime numbers, we find a recipe that, if followed step by step, solves the following problem: given two numbers, find the largest number that divides them both.

It may help to put the problem in more visual terms. Imagine that the floor of your kitchen is 36 feet long by 15 feet wide. You want to know the largest square tile that will enable you to cover the entire floor without cutting any tiles. So what should you do? Here is the 2000-year-old algorithm that solves the problem:

Suppose you have two numbers, M and N (and suppose N is smaller than M). Start by dividing M by N and call the remainder N_1. If N_1 is zero, then N is the largest number that divides them both. If N_1 is not zero, then divide N by N_1 and call the remainder N_2. If N_2 is zero, then N_1 is the largest number that divides M and N. If N_2 is not zero, then do the same thing again. Divide N_1 by N_2 and call the remainder N_3. These remainders are getting smaller and smaller and are whole numbers, so at some point one must hit zero. When it does, the algorithm guarantees that the previous remainder is the largest number that divides both M and N. This number is known as the highest common factor or greatest common divisor.

Now let's return to our challenge of tiling the kitchen floor. First we find the largest square tile that will fit inside the original shape. Then we look for the largest square tile that will fit inside the remaining part – and so on, until you hit a square tile that finally covers the remaining space evenly. This is the largest square tile that will enable you to cover the entire floor without cutting any tiles.

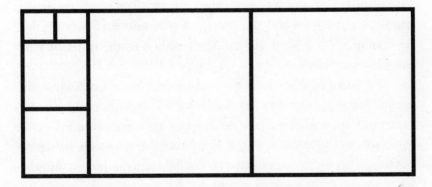

If $M = 36$ and $N = 15$, then dividing N into M gives you a remainder of $N_1 = 6$. Dividing N_1 into N we get a remainder of $N_2 = 3$. But now dividing N_1 by N_2 we get no remainder at all, so we know that 3 is the largest number that can divide both 36 and 15.

You see that there are lots of 'if . . . , then . . .' clauses in this process. That is typical of an algorithm and is what makes algorithms so perfect for coding and computers. Euclid's ancient recipe gets to the heart of four key characteristics any algorithm should ideally possess:

1. It should consist of a precisely stated and unambiguous set of instructions.
2. The procedure should always finish, regardless of the numbers you insert. (It should not enter an infinite loop!)
3. It should give the answer for any values input into the algorithm.
4. Ideally it should be fast.

In the case of Euclid's algorithm, there is no ambiguity at any stage. Because the remainder grows smaller at every step, after a finite number of steps it must hit zero, at which point the algorithm stops and spits out the answer. The bigger the numbers, the longer the algorithm will take, but it's still proportionally fast. (The number of steps is five times the number of digits in the smaller of the two numbers, for those who are curious.)

If one of the oldest algorithms is 2000 years old then why does it owe its name to a ninth-century Persian mathematician? Muhammad Al-Khwarizmi was one of the first directors of the great House of Wisdom in Baghdad and was responsible for many of the translations of the Ancient Greek mathematical texts into Arabic. 'Algorithm' is the Latin interpretation of his name. Although all the instructions for Euclid's algorithm are there in the *Elements*, the language that Euclid used was very clumsy. The Ancient Greeks thought very geometrically, so numbers were lengths of lines and proofs consisted of pictures – a bit like our example with tiling the kitchen floor. But pictures are not a rigorous way to do mathematics. For that you need the language of algebra, where a letter can stand for any number. This was the invention of Al-Khwarizmi.

To be able to articulate clearly how an algorithm works you need a language that allows you to talk about a number without specifying what that number is. We already saw it at work in explaining how Euclid's algorithm worked. We gave names to the numbers that we were trying to analyse: N and M. These variables can represent any number. The power of this new linguistic take on mathematics meant that it allowed mathematicians to understand the grammar that underlies the way that numbers work. You didn't have to talk about particular examples where the method worked. This new language of algebra provided a way to explain the patterns that lie behind the behaviour of numbers. A bit like a code for running a program, it shows why it would work whatever numbers you chose, the third criterion in our conditions for a good algorithm.

Algorithms have become the currency of our era because they are perfect fodder for computers. An algorithm exploits the pattern underlying the way we solve a problem to guide us to a solution. The computer doesn't need to think. It just follows the instructions encoded in the algorithm again and again, and, as if by magic, out pops the answer you were looking for.

Desert island algorithm

One of the most extraordinary algorithms of the modern age is the one that helps millions of us navigate the internet every day. If I were cast away on a desert island and could only take one algorithm with me, I'd probably choose the one that drives Google. (Not that it would be much use, as I'd be unlikely to have an internet connection.)

In the early days of the internet (we're talking the early 1990s) there was a directory that listed all of the existing websites. In 1994 there were only 3000 of them. The internet was small enough for you to pretty easily thumb through and find what you were looking for. Things have changed quite a bit since then. When I started writing this paragraph there were 1,267,084,131 websites live on the internet. A few sentences later that number has gone up to 1,267,085,440. (You can check the current status here: http://www.internetlivestats.com/.)

How does Google figure out exactly which one of the billion websites to recommend? Mary Ashwood, an 86-year-old granny from Wigan, was careful to send her requests with a courteous 'please' and 'thank you', perhaps imagining an industrious group of interns on the other end sifting through the endless requests. When her grandson Ben opened her laptop and found 'Please translate these roman numerals mcmxcviii thank you', he couldn't resist tweeting the world about his nan's misconception. He got a shock when someone at Google replied with the following tweet:

Dearest Ben's Nan.
Hope you're well.
In a world of billions of Searches, yours made us smile.
Oh, and it's 1998.
Thank YOU

Ben's Nan brought out the human in Google on this occasion, but there is no way any company could respond personally to the million searches Google receives every fifteen seconds. So if it isn't magic Google elves scouring the internet, how does Google succeed in so spectacularly locating the answers you want?

It all comes down to the power and beauty of the algorithm Larry Page and Sergey Brin cooked up in their dorm rooms at Stanford in 1996. They originally wanted to call their new algorithm 'Backrub', but eventually settled instead on 'Google', inspired by the mathematical number for one followed by 100 zeros, which is known as a googol. Their mission was to find a way to rank pages on the internet to help navigate this growing database, so a huge number seemed like a cool name.

It isn't that there weren't other algorithms out there being used to do the same thing, but these were pretty simple in their conception. If you wanted to find out more about 'the polite granny and Google', existing algorithms would have identified all of the pages with these words and listed them in order, putting the websites with the most occurrences of the search terms up at the top.

That's OK but easily hackable: any florist who sticks into their webpage's meta-data the words 'Mother's Day Flowers' a thousand times will shoot to the top of every son or daughter's search. You want a search engine that can't easily be pushed around by savvy web designers. So how can you come up with an unbiased measure of the importance of a website? And how can you find out which sites you can ignore?

Page and Brin struck on the clever idea that if a website has many links pointing to it, then those other sites are signalling that it is worth visiting. The idea is to democratise the measure of a website's worth by letting other websites vote for who they think is important. But, again, this could be hacked. I just need to set up a thousand artificial websites linking to my florist's website and it will bump the site up the list. To prevent this, they decided to give more weight to a vote that came from a website that itself commanded respect.

This still left them with a challenge: how do you rank the importance of one site over another? Take this mini-network (*top of page 50*) as an example.

We want to start by giving each site equal weight. Let's think of the websites as buckets; we'll give each site eight balls

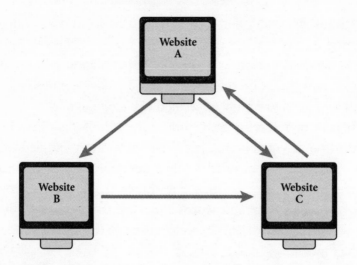

to indicate that they have equal rank. Now the websites have to give their balls to the sites they link to. If they link to more than one site, then they will share their balls equally. Since website A links to both website B and website C, for example, it will give 4 balls to each site. Website B, however, has decided to link only to website C, putting all eight of its balls into website C's bucket (*see page 51*).

After the first distribution, website C comes out looking very strong. But we need to keep repeating the process because website A will be boosted by the fact that it is being linked to by the now high-ranking website C. The table below shows how the balls move around as we iterate the process.

	Round 1	Round 2	Round 3	Round 4	Round 5	Round 6	Round 7	Round 8	Round 9
A	8	8	12	8	10	10	9	10	9.5
B	8	4	4	6	4	5	5	4.5	5
C	8	12	8	10	10	9	10	9.5	9.5

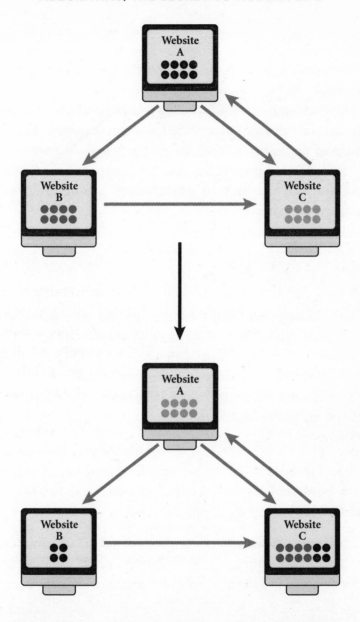

At the moment, this does not seem to be a particularly good algorithm. It appears not to stabilise and is rather inefficient, failing two of our criteria for the ideal algorithm. Page and Brin's great insight was to realise that they needed to find a way to assign the balls by looking at the connectivity of the network. It turned out they'd been taught a clever trick in their university mathematics course that worked out the correct distribution in one step.

The trick starts by constructing a matrix which records the way that the balls are redistributed among the websites. The first column of the matrix records the proportion going from website A to the other websites. In this case 0.5 goes to website B and 0.5 to website C. The matrix of redistribution is therefore given by the following matrix:

$$\begin{pmatrix} 0 & 0 & 1 \\ 0.5 & 0 & 0 \\ 0.5 & 1 & 0 \end{pmatrix}$$

The challenge is to find what is called the eigenvector of this matrix with eigenvalue 1. This is a column vector that does not get changed when multiplied by the matrix.* Finding these eigenvectors or stability points is something we teach undergraduates early on in their university career. In the case of our network we find that the following column vector is stabilised by the redistribution matrix:

$$\begin{pmatrix} 0 & 0 & 1 \\ 0.5 & 0 & 0 \\ 0.5 & 1 & 0 \end{pmatrix} \begin{pmatrix} 2 \\ 1 \\ 2 \end{pmatrix} = \begin{pmatrix} 2 \\ 1 \\ 2 \end{pmatrix}$$

This means that if we split the balls in a 2:1:2 distribution we see that this weighting is stable. Distribute the balls using our previous game and the sites still have a 2:1:2 distribution.

---

* Here is the rule for multiplying matrices:

$$\begin{pmatrix} a & b & c \\ d & e & f \\ g & h & i \end{pmatrix} \begin{pmatrix} x \\ y \\ z \end{pmatrix} = \begin{pmatrix} ax + by + cz \\ dx + ey + fz \\ gx + hy + iz \end{pmatrix}$$

Eigenvectors of matrices are an incredibly potent tool in mathematics and the sciences more generally. They are the secret to working out the energy levels of particles in quantum physics. They can tell you the stability of a rotating fluid like a spinning star or the reproduction rate of a virus. They may even be key to understanding how prime numbers are distributed throughout all numbers.

By calculating the eigenvector of the network's connectivity we see that websites A and C should be ranked equally. Although website A is linked to by only one site (website C), the fact that website C is highly valued and links only to website A means that its link bestows high value to website A.

This is the basic core of the algorithm. There are a few extra subtleties that need to be introduced to get the algorithm working in its full glory. For example, the algorithm needs to take into account anomalies like websites that don't link to any other websites and become sinks for the balls being redistributed. But at its heart is this simple idea.

Although the basic engine is very public, there are parameters inside the algorithm that are kept secret and change over time, and which make the algorithm a little harder to hack. But the fascinating thing is the robustness of the Google algorithm and its imperviousness to being gamed. It is very difficult for a website to do anything on its own site that will increase its rank. It must rely on others to boost its position. If you look at the websites that Google's page rank algorithm scores highly, you will see a lot of major news sources and university websites like Oxford and Harvard. This is because many outside websites will link to findings and opinions on university websites, because the research we do is valued by many people across the world.

Interestingly this means that when anyone with a website within the Oxford network links to an external site, the link will cause a boost to the external website's page rank, as Oxford is sharing a bit of its huge prestige (or cache of balls) with that website. This is why I often get requests to link from my website in the maths department at Oxford to external websites. The link will help increase the external website's rank and, it is hoped, make it appear on the first page of a Google search, the ultimate holy grail for any website.

But the algorithm isn't immune to clever attacks by those who understand how the mathematics works. For a short period in the summer of 2018, if you googled 'idiot' the first image that appeared was that of Donald Trump. Activists had understood how to exploit the powerful position that the website Reddit has on the internet. By getting people to vote for a post on the site containing the words 'idiot' and an image of Trump, the connection between the two shot to the top of the Google ranking. The spike was smoothed out over time by the algorithm rather than by manual intervention. Google does not like to play God but trusts in the long run in the power of its mathematics.

The internet is of course a dynamic beast, with new websites emerging every nanosecond and new links being made as existing sites are shut down or updated. This means that page ranks need to change dynamically. In order for Google to keep pace with the constant evolution of the internet, it must regularly trawl through the network and update its count of the links between sites using what it rather endearingly calls 'Google spiders'.

Tech junkies and sports coaches have discovered that this way of evaluating the nodes in a network can also be applied to other networks. One of the most intriguing external applications has been in the realm of football (of the European kind, which Americans think of as soccer). When sizing up the opposition, it can be important to identify a key player who will control the way the team plays or be the hub through which all play seems to pass. If you can identify this player and neutralise them early on in the game, then you can effectively close down the team's strategy.

Two London-based mathematicians, Javier López Peña and Hugo Touchette, both football fanatics, decided to see whether Google's algorithm might help analyse the teams gearing up for the World Cup. If you think of each player as a website and a pass from one player to another as a link from one website to another, then the passes made over the course of a game can be thought of as a network. A pass to a teammate is a mark of the trust you put in that player – players generally avoid passing to a weak team-mate who might easily lose the ball, and you will only be passed to if you make yourself available. A static player will rarely be available for a pass.

They decided to use passing data made available by FIFA during the 2010 World Cup to see which players ranked most highly. The results were fascinating. If you analysed England's style of play, two players, Steven Gerrard and Frank Lampard, emerged with a markedly higher rank than others. This reflects the fact that the ball very often went through these two mid-fielders: take them out and England's game collapses. England did not get very far that year in the World Cup – they were knocked out early by their old nemesis, Germany.

Contrast this with the eventual winners: Spain. The algorithm shared the rank uniformly around the whole team, indicating that there was no clear hub through which the game was being played. This is a reflection of the very successful 'total football' or 'tiki-taka' style played by Spain, in which players constantly pass the ball around, a strategy that contributed to Spain's ultimate success.

Unlike many sports in America that thrive on data, it has taken some time for football to take advantage of the mathematics and statistics bubbling underneath the game. But by the 2018 World Cup in Russia many teams boasted a scientist on board to crunch the numbers to understand the strengths and weaknesses of the opposition, including how the network of each team behaves.

A network analysis has even been applied to literature. Andrew Beveridge and Jie Shan took the epic saga *A Song of Ice and Fire* by George R. R. Martin, otherwise known as *Game of Thrones*. Anyone who knows the story will be aware that predicting which characters will make it through to the next volume, let alone the next chapter, is notoriously tricky, as Martin is ruthless at killing off even the best characters he has created.

Beveridge and Shan decided to create a network between characters in the books. They identified 107 key people who became the nodes of the network. The characters were then connected with weighted edges according to the strength of the relationship. But how can an algorithm assess the importance of a connection? The algorithm was simply asked to count the number of times the two names appeared in the text within fifteen words of each other. This doesn't measure friendship – it indicates some measure of interaction or connection between them.

They decided to analyse the third volume in the series, *A Storm of Swords*, as the narrative had settled down by this point, and began by constructing a page rank analysis of the nodes or characters in the network. Three characters quickly stood out as important to the plot: Tyrion, Jon Snow and Sansa Stark. Anyone who has read the books or seen the series would not be surprised by this revelation. What is striking is that a computer algorithm which does not understand what it is reading achieved this same insight. It did so not simply by counting how many times a character's name appears – that would pull out other names. It turned out that a subtler analysis of the network revealed the true protagonists.

To date, all three characters have survived Martin's ruthless pen which has cut short some of the other key characters in the third volume. This is the mark of a good algorithm: it can be used in multiple scenarios. This one can tell you something useful from football to *Game of Thrones*.

Maths, the secret to a happy marriage

Sergey Brin and Larry Page may have cracked the code to steer you to websites you don't even know you're looking for, but can an algorithm really do something as personal as find your soulmate? Visit OKCupid and you'll be greeted by a banner proudly declaring: 'We use math to find you dates'.

These dating websites use a 'matching algorithm' to search through profiles and match people up according to their likes, dislikes and personality traits. They seem to be doing a pretty good job. In fact, the algorithms seem to be better than we are on our own: recent research published in the *Proceedings of the National Academy of Sciences* looked at 19,000 people who married between 2005 and 2012 and found that those who met online were happier and had more stable marriages.

The first algorithm to win its creators a Nobel Prize, originally formulated by two mathematicians, David Gale and Lloyd Shapley, in 1962, used a matching algorithm to solve something called 'the Stable Marriage Problem'. Gale, who died in 2008, missed out on the award, but Shapley shared the prize in 2012 with the economist Alvin Roth, who saw the importance of the algorithm not just to the question of relationships but also to social problems including assigning health care and student places fairly.

Shapley was amused by the award: 'I consider myself a mathematician and the award is for economics,' he said at the time, clearly surprised by the committee's decision. 'I never, never in my life took a course in economics.' But the mathematics he cooked up has had profound economic and social implications.

The Stable Marriage Problem that Shapley solved with Gale sounds more like a parlour game than a piece of cutting-edge economic theory. To illustrate the precise nature of the problem, imagine you've got four heterosexual men and four heterosexual women. They've been asked to list the four members of the opposite sex in order of preference. The challenge for the algorithm is to match them up in such a way as to come up with stable marriages. What this means is that there shouldn't be a man and woman who would prefer to be with one another than with the partner they've been assigned. Otherwise there's a good chance that at some point they'll leave their partners to run off with one another. At first sight it isn't at all clear, even with four pairs, that it is possible to arrange this.

Let's take a particular example and explore how Gale and Shapley could guarantee a stable pairing in a systematic and algorithmic manner. The four men will be played by the kings from a pack of cards: King of Spades, King of Hearts, King of Diamonds and King of Clubs. The women are the corresponding queens. Each king and queen has listed his or her preferences:

For the kings:

	K. ♠	K. ♥	K. ♦	K. ♣
1st choice	Q. ♠	Q. ♦	Q. ♣	Q. ♦
2nd choice	Q. ♦	Q. ♣	Q. ♥	Q. ♥
3rd choice	Q. ♥	Q. ♠	Q. ♦	Q. ♠
4th choice	Q. ♣	Q. ♥	Q. ♠	Q. ♣

For the queens:

	Q. ♠	Q. ♥	Q. ♦	Q. ♣
1st choice	K. ♥	K. ♣	K. ♠	K. ♥
2nd choice	K. ♠	K. ♠	K. ♦	K. ♦
3rd choice	K. ♦	K. ♥	K. ♥	K. ♠
4th choice	K. ♣	K. ♦	K. ♣	K. ♣

Now suppose you were to start by proposing that each king be paired with the queen of the same suit. Why would this result in an unstable pairing? The Queen of Clubs has ranked the King of Clubs as her least preferred partner so frankly she'd be happier with any of the other kings. And check out the King of Hearts' list: the Queen of Hearts is at the bottom of his list. He'd certainly prefer the Queen of Clubs over the option he's been given. In this scenario, we can envision the Queen of Clubs and the King of Hearts running away together. Matching kings and queens via their suits would lead to unstable marriages.

How do we match them so we won't end up with two cards running off with each other? Here is the recipe Gale and Shapley cooked up. It consists of several rounds of proposals by the queens to the kings until a stable pairing finally emerges. In the first round of the algorithm, the queens all propose to their first choice. The Queen of Spades' first choice is the King of Hearts. The Queen of Hearts' first choice is the King of Clubs. The Queen of Diamonds chooses the King of Spades and the Queen of Clubs proposes to the King of Hearts. So it seems that the King of Hearts is the heart-throb of the pack, having received two proposals. He chooses which of the two queens he prefers, which is the Queen of Clubs, and rejects the Queen of Spades. So we have three provisional engagements, and one rejection.

First round

K. ♠	K. ♥	K. ♦	K. ♣
Q. ♦	Q. ♠ Q. ♣		Q. ♥

The rejected queen strikes off her first-choice king and in the next round moves on to propose to her second choice: the King of Spades. But now the King of Spades has two proposals. His first proposal from round one, the Queen of Diamonds, and a new proposal from the Queen of Spades. Looking at his ranking, he'd actually prefer the Queen of Spades. So he rather cruelly rejects the Queen of Diamonds (his provisional engagement on the first round of the algorithm).

Second round

K. ♠	K. ♥	K. ♦	K. ♣
Q. ♦ Q. ♠	Q. ♣		Q. ♥

Which brings us to round three. In each round, the rejected queens propose to the next king on their list and the kings always go for the best offer they receive. In this third round the rejected Queen of Diamonds proposes to the King of Diamonds (who has been standing like that kid who never gets picked for the team). Despite the fact that the Queen of Diamonds is low down on his list, he hasn't got a better option, as the other three queens prefer other kings who have accepted them.

Third round

K. ♠	K. ♥	K. ♦	K. ♣
Q. ♠	Q. ♣	Q. ♦	Q. ♥

Finally everyone is paired up and all the marriages are stable. Although we have couched the algorithm in terms of a cute parlour game with kings and queens, the algorithm is now used all over the world: in Denmark to match children to day-care places; in Hungary to match students to schools; in New York to allocate rabbis to synagogues; and in China, Germany and Spain to match students to universities. In the UK it has been used by the

National Health Service to match patients to organ donations, resulting in many lives being saved.

And it is building on top of the puzzle solved by Gale and Shapley that the modern algorithms which run our dating agencies are based. The problem is more complex since information is incomplete. Preferences are movable and relative, and shift even within relationships from day to day. But essentially the algorithms are trying to match people with preferences that will lead to a stable and happy pairing. And the evidence suggests that the algorithms could well be better than leaving it to human intuition.

You might have detected an interesting asymmetry in the algorithm that Gale and Shapley cooked up. We got the queens to propose to the kings. Would it have mattered if we had invited the kings to propose to the queens instead? Rather strikingly it does. You would end up with a different stable pairing if you applied the algorithm by swapping kings and queens.

The Queen of Diamonds would end up with the King of Hearts and the Queen of Clubs with the King of Diamonds. The two queens swap partners, but now they're paired up with slightly lower choices. Although both pairings are stable, when queens propose to kings, the queens end up with the best pairings they could hope for. Flip things around and the kings are better off.

Medical students in America looking for residencies realised that hospitals were using this algorithm to assign places in such a way that the hospitals did the proposing. This meant the students were getting a worse deal. After some campaigning by students who pointed out how unfair this was, eventually the algorithm was reversed to give students the better end of the deal.

This is a powerful reminder that, as our lives are increasingly pushed around by algorithms, it's important to understand how they work and what they're doing, because otherwise you may be getting shafted.

The battle of the booksellers

The trouble with algorithms is that sometimes there are unexpected consequences. A human might be able to tell that something weird was happening, but an algorithm will just carry on doing what it was programmed to do, regardless of how absurd the consequences may be.

My favourite example of this centres on two second-hand booksellers who ran their shops using algorithms. A postdoc working at UC Berkeley was keen to get hold of a copy of Peter Lawrence's book *The Making of a Fly*. It is a classic published in 1992 that developmental biologists often use, but by 2011 the text had been out of print for some time. The postdoc was after a second-hand copy.

Checking on Amazon, he found a number of copies priced at about $40, but then was rather shocked to see a copy on sale for $1,730,045.91. The seller, profnath, wasn't even including shipping in the bargain. Then he noticed that there was another copy on sale for even more! This seller, bordeebook, was asking a staggering $2,198,177.95 (plus $3.99 for shipping of course).

The postdoc showed this to his supervisor, Michael Eisen, who presumed it must be a graduate student having fun. But both booksellers had very high ratings and seemed to be legitimate. Profnath had had over 8000 recommendations over the last twelve months, while bordeebook had had over 125,000 during the same period. Perhaps it was just a weird blip.

When Eisen checked the next day to see if the prices had dropped to more sensible levels, he found instead that they'd gone up. Profnath now wanted $2,194,443.04 while bordeebook was asking a phenomenal $2,788,233.00. Eisen decided to put his scientific hat on and analyse the data. Over the next few days he tracked the changes in an effort to work out if there was some pattern to the strange prices.

	profnath	bordeebook	profnath/ bordeebook	bordeebook/ profnath
8 April	$1,730,045.91	$2,198,177.95		1.27059
9 April	$2,194,443.04	$2,788,233.00	0.99830	1.27059
10 April	$2,783,494.00	$3,536,675.57	0.99830	1.27059
11 April	$3,530,663.65	$4,486,021.69	0.99830	1.27059
12 April	$4,478,395.76	$5,690,199.43	0.99830	1.27059
13 April	$5,680,526.66	$7,217,612.38	0.99830	1.27059

Eventually he spotted the mathematical rule behind the escalating prices. Divide the profnath price by the bordeebook price from the day before and you always got 0.99830. Divide the bordeebook price by the profnath book on the same day and you always got 1.27059. Each seller had programmed their website to use an algorithm that was setting the prices for books they were selling. Each day the profnath algorithm would check the price of the book at bordeebook and would then multiply it by 0.99830. This algorithm made perfect sense because the seller was programming the site to slightly undercut the competition at bordeebook. It is the algorithm at bordeebook that is slightly more curious. It was programmed to detect any price change in its rival and to multiply this new price by a factor of 1.27059.

The combined effect was that each day the price would be multiplied by 0.99830 x 1.27059, or 1.26843. This ensured that the price would grow exponentially. If profnath had set a sharper factor to undercut the price being offered by bordeebook, you would have seen the price collapse over time rather than escalate.

The explanation for profnath's algorithm seems clear, but why was bordeebook's algorithm set to offer the book at a higher price? Surely no one would buy the more expensive book? Perhaps they were relying on their bigger reputation with a greater number of positive recommendations to drive traffic their way, especially if their price was only slightly higher, which at the start it would have been. As Eisen wrote in his blog, 'this seems like a fairly risky thing to rely on. Meanwhile you've got a book sitting on the shelf collecting dust. Unless, of course, you don't actually have the book . . .'

Then the truth suddenly dawned on him. Of course. They didn't actually have the book! The algorithm was programmed to see what books were out there and to offer the same book at a slight markup. If someone wanted the book from their reliable bordeebook's website, then bordeebook would go and purchase it from the other bookseller and sell it on. But to cover costs this would necessitate a bit of a markup. The algorithm thus multiplied the price by a factor of 1.27059 to cover the purchase of the book, the shipping and a little extra profit.

Using a few logarithms it's possible to work out that the book most likely first went on sale forty-five days before 8 April at about $40. This shows the power of exponential growth. It only took a month and a half for the price to reach into the millions! The price peaked at $23,698,655.93 (plus $3.99 shipping) on 18 April, when finally a human at profnath intervened, realising that something strange was going on. The price then dropped to $106.23. Predictably bordeebook's algorithm offered their book at $106.23 x 1.27059 = $134.97.

The mispricing of *The Making of a Fly* did not have a devastating impact for anyone involved, but there are more serious cases of algorithms used to price stock options causing flash crashes on the markets. The unintended consequences of algorithms is one of the prime sources of the existential fears people have about advancing technology. What if a company builds an algorithm that is tasked with maximising the collection of carbon, but it suddenly realises the humans who work in the factory are carbon-based organisms, so it starts harvesting the humans in the factory for carbon production? Who would stop it?

Algorithms are based on mathematics. At some level they are mathematics in action. But they don't really creatively stretch the field. No one in the mathematical community feels particularly threatened by them. We don't really believe that algorithms will turn on their creators and put us out of a job. For years I believed that these algorithms would do no more than speed up the mundane part of my work. They were just more sophisticated versions of Babbage's calculating machine that could be told to do the algebraic or numerical manipulations which would take me tedious hours to write out by hand. I always felt in control. But that is all about to change.

Up till a few years ago it was felt that humans understood what their algorithms were doing and how they were doing it. Like Lovelace, they believed you couldn't really get more out than you put in. But then a new sort of algorithm began to emerge, an algorithm that could adapt and change as it interacted with its data. After a while its programmer may not understand quite why it is making the choices it is. These programs were starting to produce surprises, and for once you could get more out than you put in. They were beginning to be more creative. These were the algorithms DeepMind exploited in its crushing of humanity in the game of Go. They ushered in the new age of machine learning.

5

FROM TOP DOWN TO BOTTOM UP

Machines take me by surprise with great frequency.
Alan Turing

I first met Demis Hassabis a few years before his great Go triumph at a meeting about the future of innovation. New companies were on the lookout for investment from venture capitalists and investors. Some were going to transform the future, but most would flash and burn. The art was for VCs and angel investors to spot the winners. I must admit when I heard Hassabis speak about code that could learn, adapt and improve I dismissed him out of hand. I couldn't see how, if you were programming a computer to play a game, the program could get any further than the person who was writing the code. How could you get more out than you were putting in? I wasn't the only one. Hassabis admits that getting investors to give money to AI a decade ago was extremely difficult.

How I wish now that I'd backed that horse as it came trotting by! The transformative impact of the ideas Hassabis was proposing can be judged by the title of a recent session on AI: 'Is machine learning the new 42?' (The allusion to Douglas Adams's answer to the question of life, the universe and everything from his book *The Hitchhiker's Guide to the Galaxy* would have been familiar to the geeky attendees, many of whom were brought up on a diet of sci-fi.) So what has happened to spark this new AI revolution?

The simple answer is data. It is an extraordinary fact that 90 per cent of the world's data has been created in the last five years. 1 exabyte (10^{18}) of data is created on the internet every day, roughly the equivalent of the amount of data that can be stored on 250 million DVDs. Humankind now produces in two days the same amount of data it took us from the dawn of civilisation until 2003 to generate.

This flood of data is the main catalyst for the new age of machine learning. Before now there just wasn't enough of an environment for an algorithm to roam around in and learn. It was like having a child and denying it sensory input. We know that children who have been trapped indoors fail to develop language and other basic skills. Their brains may have been primed to learn but didn't encounter enough stimulus or experience to develop properly.

The importance of data to this new revolution has led many to speak of data as the new oil. If you have access to data you are straddling the twenty-first-century's oilfields. This is why the likes of Facebook, Twitter, Google and Amazon are sitting pretty – we are giving them our reserves for free. Well, not exactly for free as we are exchanging our data for the services they provide. When I drive in my car using Waze, I have chosen to exchange data about my location in return for the most efficient route to my destination. The trouble is, many people are not aware of these transactions and give up valuable data for little in return.

At the heart of machine learning is the idea that an algorithm can be created that will find new questions to ask if it gets something wrong. It learns from its mistake. This tweaks the algorithm's equations such that next time it will act differently and won't make the same mistake. This is why access to data is so important: the more examples these smart algorithms can train on the more experienced they will become, and the more each tweak will refine them. Programmers are essentially creating a meta-algorithm which creates new algorithms based on the data it encounters.

People in the field of AI have been shocked at the effectiveness of this new approach. Partly this is because the underlying technology is not that new. These algorithms are created by building up layers of questions that can help reach a conclusion. These layers are sometimes called neural networks because they mimic the way the human brain works. If you think about the structure of the brain, neurons are connected to other neurons by synapses. A collection of neurons might fire due to an input of data from our senses. (The smell of freshly baked bread.) Secondary neurons will then fire, provided certain thresholds are passed. (The decision to eat the bread.) A secondary neuron might fire if ten connected neurons are firing due to the input data, for instance, but not if fewer are firing. The trigger might depend also on the strength of the incoming signal from the other neurons.

Already in the 1950s computer scientists created an artificial version of this process, which they called the perceptron. The idea is that a neuron is like a logic gate that receives input and then, depending on a calculation, decides either to fire or not.

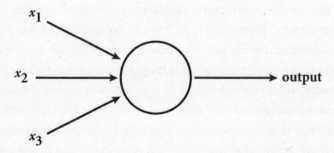

Let's imagine that the perceptron receives three input numbers. It weights the importance of each of these. In the diagram on page 68, perhaps x_1 is three times as important as x_2 and x_3. It would calculate $3x_1 + x_2 + x_3$ and then, depending on whether this fell above or below a certain threshold, it would fire an output or not. Machine learning hinges on reweighting the input if it gets the answer wrong. For example, perhaps x_3 is more important in making a decision than x_2, so you might change the equation to $3x_1 + x_2 + 2x_3$. Or perhaps we simply need to tweak the activation level so the threshold can be dialled up or down in order to fire the perceptron. We can also create a perceptron such that the degree to which it fires is proportional to by how much the function has passed the threshold. The output can be a measure of its confidence in the assessment of the data.

Let's cook up a perceptron to decide whether you are going to go out tonight. It will depend on three things: (1) is there anything good on TV; (2) are your friends going out; (3) what night of the week is it? Give each of these variables a score between 0 and 10, to indicate your level of preference. For example, Monday will get a 1 score while Friday will get a 10. Depending on your personal proclivities, some of these variables might count more than others. Perhaps you are a bit of a couch potato, so anything vaguely decent on TV will cause you to stay in. This would mean that the x_1 variable scores high. The art of this equation is tuning the weightings and the threshold value to mimic the way you behave.

Just as the brain consists of a whole chain of neurons, perceptrons can be layered, so that the triggering of nodes gradually causes a cascade through the network. This is what we call a neural network. In fact, there is a slightly subtler version of the perceptron called the sigmoid neuron that smoothes out the behaviour of these neurons so that they aren't just simple on/off switches.

Given that computer scientists had already understood how to create artificial neurons, why did it take so long to make these things work so effectively? This brings us back to data. The perceptrons need data from which to learn and evolve; together

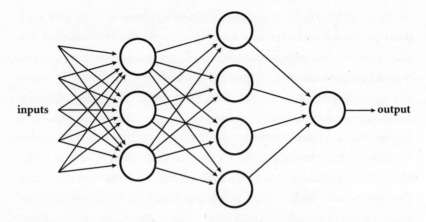

these are the two ingredients you need to create an effective algorithm. We could try to program our perceptron to decide when we should go out by assigning weights and thresholds, but it is only by training it on our actual behaviour that it will have any chance of getting it right. Each failure to predict our behaviour allows it to learn and reweight itself.

To see or not to see

One of the big hurdles for AI has always been computer vision. Five years ago computers were terrible at understanding what it was they were looking at. This is one domain where the human brain totally outstrips its silicon rivals. We are able to eyeball a picture very quickly and say what it is or to classify different regions of the image. A computer could analyse millions of pixels, but programmers found it very difficult to write an algorithm that could take all this data and make sense of it. How can you create an algorithm from the top down to identify a cat? Each image consists of a completely different arrangement of pixels and yet the human brain has an amazing ability to synthesise this data and integrate the input to output the answer, 'cat'.

This ability of the human brain to recognise images has been used to create an extra layer of security at banks, and to make sure you aren't a robot trawling for tickets online. In essence you needed to pass an inverse Turing Test. Shown an image or some strange handwriting, humans are very good at saying what the image or script is. Computers couldn't cope with all the variations. But machine learning has changed all that.

Now, by training on data consisting of images of cats, the algorithm gradually builds up a hierarchy of questions it can ask an image that, with a high probability of accuracy, will identify it as a cat. These algorithms are slightly different in flavour to those we saw in the last chapter, and violate one of the four conditions we put forward for a good algorithm. They don't work 100 per cent of the time. But they do work most of the time. The point is to get that 'most' as high as possible. The move from deterministic foolproof algorithms to probabilistic ones has been a significant psychological shift for those working in the industry. It's a bit like moving from the mindset of the mathematician to that of the engineer.

You may wonder why, if this is the case, you are still being asked to identify bits of images when you want to buy tickets to the latest gig to prove you are human. What you are actually doing is helping to prepare the training data that will then be fed to the algorithms so that they can try to learn to do what you do so effortlessly. Algorithms need labelled data to learn from. What we are really doing is training the algorithms in visual recognition.

This training data is used to learn the best sorts of questions to ask to distinguish cats from non-cats. Every time it gets it wrong, the algorithm is altered so that the next time it will get it right. This might mean altering the parameters of the current algorithm or introducing a new feature to distinguish the image more accurately. The change isn't communicated in a top-down manner by a programmer who is thinking up all of the questions in advance.

The algorithm builds itself from the bottom up by interacting with more and more data.

I saw the power of this bottom-up learning process at work when I dropped in to the Microsoft labs in Cambridge to see how the Xbox which my kids use at home is able to identify what they're doing in front of the camera as they move about. This algorithm has been created to distinguish hands from heads, and feet from elbows. The Xbox has a depth-sensing camera called Kinect which uses infrared technology to record how far obstacles are from the camera. If you stand in front of the camera in your living room it will detect that your body is nearer than the wall at the back of the room and will also be able to determine the contours of your body.

But people come in different shapes and sizes. They can be in strange positions, especially when playing Xbox. The challenge for the computer is to identify thirty-one distinct body parts, from your left knee to your right shoulder. Microsoft's algorithm is able to do this on a single frozen image. It does not use the way you are moving (which requires more processing power to analyse and would slow the game down).

So how does it manage to do this? The algorithm has to decide for each pixel in each image which of the thirty-one body parts it belongs to. Essentially it plays a game of twenty questions. In fact, there's a sneaky algorithm you can write for the game of twenty questions that will guarantee you get the right answer. First ask: 'Is the word in the first half of the dictionary or the second?' Then narrow down the region of the dictionary even more by asking: 'Is it in the first or second half of the half you've just identified?' After twenty questions this strategy divides the dictionary up into 2^{20} different regions. Here we see the power of doubling. That's more than a million compartments – far more than there are entries in the *Oxford English Dictionary*, which roughly come to 300,000.

But what questions should we ask our pixels if we want to

identify which body part they belong to? In the past we would have had to come up with a clever sequence of questions to solve this. But what if we programmed the computer so that it finds the best questions to ask? By interacting with more and more data – more and more images – it finds the set of questions that seem to work best. This is machine learning at work.

We have to start with some candidate questions that we think might solve this problem so this isn't completely tabula rasa learning. The learning comes from refining our ideas into an effective strategy. So what sort of questions do you think might help us distinguish your arm from the top of your head?

Let's call the pixel we're trying to identify X. The computer knows the depth of each pixel, or how far away it is from the camera. The clever strategy the Microsoft team came up with was to ask questions of the surrounding pixels. For example, if X is a pixel on the top of my head, then if we look at the pixels north of pixel X they are much more likely not to be on my body and thus to have more depth. If we take pixels immediately south of X, they'll be pixels on my face and will have a similar depth. But if the pixel is on my arm and my arm is outstretched, there will be one axis, along the length of the arm, along which the depth will be relatively unchanged, but if you move out ninety degrees from this direction it quickly pushes you off the body and onto the back wall. Asking about the depth of surrounding pixels could cumulatively build up to give you an idea of the body part that pixel belongs to.

This cumulative questioning can be thought of as building a decision tree. Each subsequent question produces another branch of the tree. The algorithm starts by choosing a series of arbitrary directions to head out from and some arbitrary depth threshold: for example, head north; if the difference in depth is less than y, go to the left branch of the decision tree; if it is greater, go right – and so on. We want to find questions that give us new information. Having started with an initial random set of questions, once we apply these questions to 10,000 labelled images we start getting somewhere. (We know, for instance, that pixel X in image 872 is an elbow, and in image 3339 it is part of the left foot.) We can think of each branch or body part as a separate bucket. We want the questions to ensure that all the images where pixel X is an elbow have gone into one bucket. That is unlikely to happen on the first random set of questions. But over time, as the algorithm starts refining the angles and the depth thresholds, it will get a better sorting of the pixels in each bucket.

By iterating this process, the algorithm alters the values, moving in the direction that does a better job at distinguishing the pixels. The key is to remember that we are not looking for perfection here. If a bucket ends up with 990 out of 1000 images in which pixel X is an elbow, then that means that in 99 per cent of cases it is identifying the right feature.

By the time the algorithm has found the best set of questions, the programmers haven't really got a clue how it has come to this conclusion. They can look at any point in the tree and see the question it is asking before and after, but there are over a million different questions being asked across the tree, each one slightly different. It is difficult to reverse-engineer why the algorithm ultimately settled on this question to ask at this point in the decision tree.

Imagine trying to program something like this by hand. You'd have to come up with over a million different questions. This prospect would defeat even the most intrepid coder, but a computer is quite happy to sort through these kinds of numbers. The amazing thing is that it works so well. It took a certain creativity for the programming team to believe that questioning the depth of neighbouring pixels would be enough to tell you what body part you were looking at – but after that the creativity belonged to the machine.

One of the challenges of machine learning is something called 'over-fitting'. It's always possible to come up with enough questions to distinguish an image using the training data, but you want to come up with a program that isn't too tailored to the data it has been trained on. It needs to be able to learn something more widely applicable from that data. Let's say you were trying to come up with a set of questions to identify citizens and were given 1000 people's names and their passport numbers. 'Is your passport number 834765489?' you might ask. 'Then you must be Ada Lovelace.' This would work for the data set on hand, but it would singularly fail for anyone outside this group, as no new citizen would have that passport number.

Given ten points on a graph, it is possible to come up with an equation that creates a curve which passes through all the points. You just need an equation with ten terms. But, again, this has not really revealed an underlying pattern in the data that could be useful for understanding new data points. You want an equation with fewer terms, to avoid this over-fitting.

Over-fitting can make you miss overarching trends by inviting you to model too much detail, resulting in some bizarre predictions. Here is a graph of twelve data points for population values in the US since the beginning of the last century. The overall trend is best described by a quadratic equation, but what if we

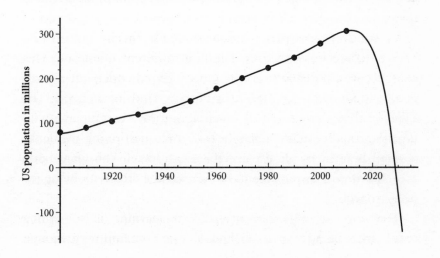

used an equation with higher powers of x than simply x^2? Taking an equation with powers all the way up to x^{11} actually gives a very tight fit to the data, but extend this equation into the future and it takes a dramatic lurch downwards, predicting complete annihilation of the US population in the middle of October in 2028. Or perhaps the maths knows something we don't!

Algorithmic hallucinations

Advances in computer vision over the last five years have surprised everyone. And it's not just the human body that new algorithms can navigate. To match the ability of the human brain to decode visual images has been a significant hurdle for any computer claiming to compete with human intelligence. A digital camera can take an image with a level of detail that far exceeds the human brain's storage capacity, but that doesn't mean it can turn millions of pixels into one coherent story. The way the brain can process data and integrate it into a narrative is something we are far from understanding, let alone replicating in our silicon friends.

Why is it that when we receive the information that comes in through our senses we can condense it into an integrated experience? We don't experience the redness of a die and its cubeness as two different experiences. They are fused into a single experience. Replicating this fusion has been one of the challenges in getting a computer to interpret an image. Reading an image one pixel at a time won't tell us much about the overall picture. To illustrate this more immediately, take a piece of paper and make a small hole in it. Now place the paper on an A4 image of a face. It's almost impossible to tell whose face it is by moving the hole around.

Five years ago this challenge still seemed impossible. But that was before the advent of machine learning. Computer programmers in the past would try to create a top-down algorithm to recognise visual images. But coming up with an 'if . . . , then . . .' set to identify an image never worked. The bottom-up strategy, allowing the algorithm to create its own decision tree based on training data, has changed everything. The new ingredient which has made this possible is the amount of labelled visual data there is now on the web. Every Instagram picture with our comments attached provides useful data to speed up the learning.

You can test the power of these algorithms by uploading an image to Google's vision website: https://cloud.google.com/vision/. Last year I uploaded an image of our Christmas tree and it came back with 97 per cent certainty that it was looking at a picture of a Christmas tree. This may not seem particularly earth-shattering, but it is actually very impressive. Yet it is not foolproof. After the initial wave of excitement has come the kick-back of limitations. Take, for instance, the algorithms that are now being trialled by the British Metropolitan Police to pick up images of child pornography online. At the moment they are getting very confused by images of deserts.

'Sometimes it comes up with a desert and it thinks it's an indecent image or pornography,' Mark Stokes, the department's head of digital and electronics forensics, admitted in a recent interview. 'For some reason, lots of people have screen-savers of deserts and it picks it up, thinking it is skin colour.' The contours of the dunes also seem to correspond to shapes the algorithms pick up as curvaceous naked body parts.

There have been many colourful demonstrations of the strange ways in which computer vision can be hacked to make the algorithm think it's seeing something that isn't there. LabSix, an independent student-run AI research group composed of MIT graduates and undergraduates, managed to confuse vision recognition algorithms into thinking that a 3D model of a turtle was in fact a gun. It didn't matter at what angle you held the turtle – you could even put it in an environment in which you'd expect to see turtles and not guns.

The way they tricked the algorithm was by layering a texture on top of the turtle that to the human eye appeared to be turtle shell and skin but was cleverly built out of images of rifles. The images of the rifle are gradually changed over and over again until a human can't see the rifle any more. The computer, however, still discerns the information about the rifle even when they are perturbed, and this ranks higher in its attempts to classify the object than the turtle on which it is printed. Algorithms have also been tricked into interpreting an image of a cat as a plate of guacamole, but LabSix's contribution is that it doesn't matter at what angle you showed the turtle, the algorithm will always be convinced it is looking at a rifle.

The same team has also shown that an image of a dog that gradually transforms pixel by pixel into two skiers on the slopes will still be classified as a dog even when the dog had completely disappeared from the screen. Their hack was all the more impressive, given that the algorithm being used was a complete black box to the hackers. They didn't know how the image was being decoded but still managed to fool the algorithm.

Researchers at Google went one step further and created images that are so interesting to the algorithm that it will ignore whatever else is in the picture, exploiting the fact that algorithms prioritise pixels they regard as important to classifying the image. If an algorithm is trying to recognise a face, it will ignore most of the background pixels: the sky, the grass, the trees, etc. The Google team created psychedelic patches of colour that totally took over and hijacked the algorithm so that while it could generally recognise a picture of a banana, when the psychedelic patch was introduced the banana disappeared from its sight. These patches can be made to register as arbitrary images, like a toaster. Whatever picture the algorithm is shown, once the patch is introduced it will think it is seeing a toaster. It's a bit like the way a dog can become totally distracted by a ball until everything else disappears from its conscious world and all it can see and think is 'ball'. Most previous attacks needed to know something about the image it was trying to misclassify, but this new patch had the virtue of working regardless of the image it was seeking to disrupt.

Humans are not tricked by these hacks, but that's not to say we're immune from similar effects. Magicians rely on the brain's tendency to be distracted by one thing in its visual field and to completely overlook what they're doing at the same time. A classic example of this is the famous video of two teams passing basketballs. If viewers are asked to count the number of passes made by one team, focusing their attention on the ball, most will completely miss the man in the monkey suit who walks through the players, bangs his chest and then walks off. These attacks on computer vision are just teasing out the algorithms' blind spots – but we have plenty of them too.

Given that driverless cars use vision algorithms to steer, it is clearly an issue that the algorithms can be attacked in this way. Imagine a stop sign that had a psychedelic sticker put on it, or a security system run by an algorithm that completely misses a gun because it thinks it's a turtle.

I decided to put the Kinect algorithm through its paces to see if I could confuse it with strange contortions of my body, but it wasn't easily fooled. Even when I did strange yoga positions that the Kinect hadn't seen in its training data it was able to identify the bits of my body with a high degree of accuracy. Since bodies out there won't be doing drastically new things, the algorithm is largely frozen and won't evolve any further. There is no need for it to keep on changing, as it is doing effectively what it was built to do. But other algorithms may need to keep adapting to new insights and changes in their environment. The algorithms that recommend the films we may like to watch, the books we may want to read, the music we may want to listen to, will have to be nimble enough to react to our changing tastes and to the stream of new creative output generated by our human code.

This is where the power of an algorithm that can continue to learn, mutate and adapt to new data comes into its own. Machine learning has opened up the prospect of algorithms that change and mature as we do.

6

ALGORITHMIC EVOLUTION

Knowledge rests not upon truth alone,
but upon error also.
Carl Jung

Today's algorithms are continually learning. This is particularly true of the recommender algorithms we trust to curate our viewing, reading and listening. As new users interact with them and convey their preferences, these algorithms have new data to learn from that helps refine their recommendations for the next user. I was intrigued to try one of these algorithms out to see how well it might know my tastes. So while I was at the Microsoft labs in Cambridge checking out the Xbox algorithm for the Kinect, I dropped in to a colleague to see one of the recommender algorithms learning in real time.

The graphic interface presented to me consisted of some 200 films randomly arranged on the screen. If I liked a film I was told to drag it to the right of the screen. I spotted a few films I enjoyed. I'm a big Wes Anderson fan so I pulled *Rushmore* over to the right. Immediately the films began to rearrange themselves on the screen. Some drifted to the right: these were other films the algorithm thought I'd be likely to enjoy. Films I might not like drifted to the left. One film isn't much to go on, so most were still clumped in the undecided middle.

I could see a film I really dislike: I find Austin Powers very annoying so I dragged that to the reject pile on the left. Again this gave the program more to go on and the films drifted further left and right, indicating that it had more confidence in its suggestion. Woody Allen's *Manhattan* was now suggested as a film I'd enjoy. I confirmed this, which didn't cause much of a ripple in the suggestions. But then I saw that it thought I would be a fan of *This is Spinal Tap*. This film had drifted quite far to the right. But I can't stand that film. So I dragged it from the right into the reject pile on the left of the screen.

Because the algorithm was expecting me to like *Spinal Tap*, it learned a lot from my telling it I didn't. The films dramatically rearranged themselves on the screen to take account of the new information. But then something more subtle happened in the back engine driving the algorithm. It learned something new from the data I had given it. This new insight altered very slightly the parameters of the model doing the recommending. The probability that I'd like *Spinal Tap* was considered too high, and so the parameters were altered in order to lower this probability. It had learned from other fans of Wes Anderson and *Manhattan* that they quite often did enjoy this film, but now it had discovered this certainly wasn't universal.

It's in this way that our interaction with dynamic algorithms allows the machines to continue learning and adapting to our likes and dislikes. These sorts of algorithms are responsible for a huge number of the choices we now make in our lives: from movies to music, from books to potential partners.

'If you like this . . .'

The basic idea of a movie recommender algorithm is quite simple. If you like films A, B and C, and another user has also indicated that these are among their favourites but also likes film D, then there is a good chance that you will also like film D. Of course, the data is much more complex than such a simple matching. You might have been drawn to films A, B and C because they had a particular actor in them who doesn't appear in film D, while the other user enjoyed them because they were all spy thrillers.

An algorithm needs to be able to look at the data and to discern why you are liking certain films. It then matches you with users who have also picked out the traits you enjoy. As with much of machine learning, you need to start with a good swathe of data. One important component of machine learning is that humans have to be used to classify the data so that computers know what it is they're looking at. This act of curating the data prepares the field in advance so that the algorithm can then pick up the underlying patterns.

With a database of movies you could ask someone to go through and pick out key characteristics – for example, identifying romcoms or sci-fi movies or perhaps movies with particular actors or directors. But this kind of curation is not ideal. It's time-consuming. It is open to biases coming from those classifying and will end up teaching the computer what we know already rather than finding new underlying trends. It could cause the algorithm to get stuck in a particularly human way of looking at the data. The best scenario is to train the algorithm to learn and spot patterns from pure raw data.

This is what Netflix was hoping to do when it issued its Netflix prize challenge in 2006. It had developed its own algorithm to push users towards films they would like but thought a competition might stimulate the discovery of better algorithms. By that point Netflix had a huge amount of data from users who had watched films and rated them on a scale of 1–5. So it decided to publish 100,480,507 ratings spanning 480,189 anonymous customers evaluating 17,770 movies. The added challenge was that these 17,770 movies were not identified. They were just given a number. There was no way to know whether film 2666 was *Bladerunner* or *Annie Hall*. All you had access to were the ratings any one of the 480,189 customers had given the film, if they had rated it at all.

In addition to the 100 million ratings that were made public, Netflix retained 2,817,131 undisclosed ratings. The challenge was to produce an algorithm that was 10 per cent better than Netflix's own algorithm in predicting what these 2,817,131 recommendations were. Given the data you had seen, your algorithm needed to predict how user 234,654 rated film 2666. To add some spice to the challenge, the first team that beat the Netflix algorithm by 10 per cent would receive a $1,000,000 prize. The catch was that if you won, you had to disclose your algorithm and give Netflix a non-exclusive licence to use the algorithm to recommend films to its users.

Several progress prizes were offered on the way to the one-million-dollar prize. Each year a prize of $50,000 would be awarded to the team that had produced the best results so far, provided they at least improved on the previous year's progress winner by 1 per cent. Again, to claim the prize you had to disclose the code you were using to drive your algorithm.

You might think it would be almost impossible to glean anything from the data, given that you haven't a clue if film 2666 was sci-fi or a comedy. But it is amazing how much even raw data gives away about itself. Think of each user as a point in 17,770-dimensional space, with one dimension for each movie, where the point moves more in one particular dimension according to whether the user has rated the film highly. Now, unless you are a mathematician, thinking about users as points in 17,770-dimensional space is rather spacey. But it's really just an extension of how you would graphically represent users if there were only three films being rated.

Suppose film 1 were *The Lion King*, film 2 *The Shining* and film 3 *Manhattan*. If a user had rated these films respectively as one star, four stars and five stars, you could imagine putting this user at the location (1,4,5) on a three-dimensional grid where the x-axis measures how much you like film 1, the y-axis film 2 and the z-axis film 3.

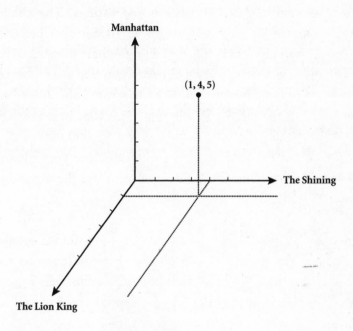

Although we can't draw a picture to represent users in 17,770-dimensional space, the mathematics can be used to plot their position throughout this space. Similarly, a film can be thought of as a point in 480,189-dimensional space, where the film moves in the dimension corresponding to a user if they have rated the film highly. At the moment it is difficult to see any patterns in all these points spread out over these massive dimensional spaces. What you want your algorithm to do is to figure out, among these points, whether there are ways to collapse the spaces to much smaller dimensions so that patterns will begin to emerge.

It's a bit like the different shadows you can see of someone's head. Some shadows reveal much more about the person than others. A profile of Hitchcock for example is very recognisable, while a shadow cast from shining a torch face-on gives away little. The idea is that films and users are like points on the face. A shadow cast at one angle might see all these points lining up, while from another angle no pattern is evident.

Perhaps you can find a way to take a two-dimensional shadow of these spaces such that the way in which users and films are mapped into the shadow, users end up next to films they are likely to enjoy. The art is finding the right shadow to reveal underlying traits that films and users might possess. Here is an example of such a shadow, created using 100 users and 500 films from the Netflix data. You can see that it is well chosen because the two

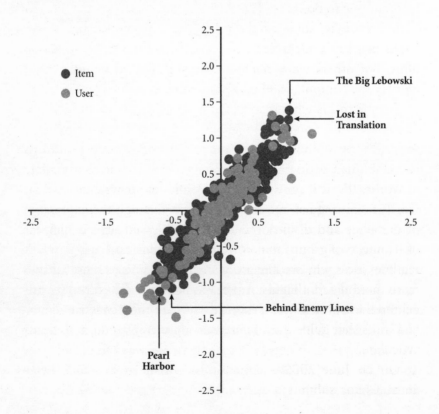

traits it is measuring seem to be quite distinct. This is borne out by the dots not being all over the place. The shadow has picked up a pattern in the data.

If you look at the actual films that were plotted, then indeed you will see that this shadow has picked out traits that we would recognise as distinct in films. Drama films are appearing in the top-right quadrant, action movies in the bottom left.

This is the approach the team that eventually won the Netflix prize in 2009 successfully implemented. They essentially sought to identify a shadow in twenty dimensions that corresponds to twenty independent traits of films that would help predict what films users would like. The power of a computer is that it can run over a whole range of different shadows and pick out the best one to reveal structure, something that our brains and eyes cannot hope to do.

Interestingly, some of the traits that the model picked out could be clearly identified: for example, action films or drama films. But others were much subtler and had no obvious label, and yet the computer had picked up a trend in the data.

For me this is what is so exciting about these new algorithms. They have the potential to tell us something new about ourselves. In a way the deep-learning algorithm is picking up traits in our human code that we still haven't been able to articulate in words. It's as if we didn't know what colour was and had no words to distinguish red from blue, but through the expression of our likes and dislikes the algorithm divided objects in front of us into two groups that correspond to blue and red. We can't really express why we like a certain movie because there are too many parameters that determined those likes. The human code behind these preferences is hidden. The computer code has identified traits that guide our preferences which we can intuit but not articulate.

On 26 June 2009 a team by the name of BellKor's Pragmatic Chaos submitted an entry which passed the 10 per cent threshold, scoring 10.05 per cent. Netflix had divided the hidden data into two halves. One half was used to give each team their score. The other half was kept back to judge the eventual winner. Once the 10 per cent threshold had been passed, other teams had one month to try to improve their score. On 25 July team Ensemble submitted an entry that scored 10.09 per cent. The next day Netflix stopped gathering new submissions. At this point both teams had upped their algorithms: BellKor's Pragmatic Chaos had hit 10.09 per cent but Ensemble had edged forward to 10.1 per cent. The prize would go to the team that scored best on the second half of the data. The result: both teams scored the same, but because BellKor's Pragmatic Chaos submitted their entry twenty minutes earlier, they walked away with one million dollars.

Given the success of this first competition, Netflix had hoped to run a second to stimulate even more innovative ideas, but it ran into a problem. The data was meant to be anonymous. Netflix had posted on the competition site the following comment about the privacy of the data:

All customer identifying information has been removed;
all that remains are ratings and dates. This follows
our privacy policy. Even if, for example, you knew all
your own ratings and their dates you probably couldn't
identify them reliably in the data because only a small
sample was included (less than one-tenth of our complete
dataset) and that data was subject to perturbation. Of
course, since you know all your own ratings that really
isn't a privacy problem is it?

Two researchers from the University of Texas at Austin were able to take the data and, by comparing it with people who had rated films on another website, the Internet Movie Database, worked out the identities of several of the users.

On 17 December 2009 four users brought a legal case against Netflix, claiming that the company had violated the video privacy protection act by releasing the data. One of the users said she was a closet lesbian mother and that the data about her movie preferences could have revealed this fact. That you might be able to infer sexual orientation or political leanings from your movie preferences has led to this being called the *Brokeback Mountain* factor. Eventually the case was settled out of court, but it led to Netflix cancelling the second round of the competition.

Data is the new oil, but we are spilling it all over the internet. Who owns that data and what can be done with it is going to be a major question for society as we head into a future fuelled by this oil.

How to train your algorithm

You may feel there is something scary about algorithms telling you what you might like if it means you will never see things it thinks you won't like. Personally, I really enjoy being directed towards new music that I might not have found by myself. I can quickly get stuck in a rut where I put on the same songs over and over. That's why I've always enjoyed the radio. But the algorithms that are now pushing and pulling me through the music library are perfectly suited to finding the gems that I'll like. My worry originally about such algorithms was that they might force everyone into certain regions of the library, leaving others bare of listeners. Would they cause a convergence of tastes? But because very often the mathematics behind them is non-linear and chaotic, this doesn't happen. A small divergence in my likes compared to yours can send us off in different directions in the library.

I listen to a lot of the music that my algorithm recommends when I am out running. It's a great place to navigate the new.

But I made a big mistake a few weeks ago. My wife asked me to help put together a playlist for her birthday party. She wanted dancing. She wanted the eighties. So we spent a couple of evenings listening to what she liked. It's not my choice of music, but we put together a great list of songs that got all our guests up and dancing. The problems came when I went out for my first run following the party. My algorithm took me into a part of the library that was full of dance music from the 1980s. I pressed skip as I ran on, but I couldn't find my way out. It took several weeks of retraining the algorithm on Shostakovich and Messiaen before I found myself back on track.

Another context in which we teach the algorithm we interact with is the spam filters on our email applications. It begins by training on a whole swathe of emails, some marked as spam, the rest considered legitimate. These are emails that aren't particular to you yet. By analysing the words that appear in these emails it starts to build up a profile of spam emails: 100 per cent of the emails that had the word 'Viagra' were spam; 99 per cent of the emails with the word 'refinance' were spam; 100 per cent of the emails with the combination 'hot Russian' were spam. The word 'diabetes' was more problematic. There seem to be a lot of spam emails going round for cures for diabetes, but it is also a word that crops up legitimately in some emails. The algorithm simply counts the split in its training data. Perhaps one in twenty emails with the word diabetes are OK, so it scores an email with 'diabetes' as having a 95 per cent chance of being spam.

Your email filter can be set at different levels of filtering. You might say that only if it's 95 per cent sure should an email go into the junk folder. But now comes the cool bit. It's trained on a generic set of emails but your actions will teach it to recognise the sorts of things you are interested in. It will learn to be responsive to the sorts of emails you are sent. Suppose that in fact you suffer from diabetes. At first all emails with the word 'diabetes' will go into your junk folder. But gradually, as you mark more and more emails including the word 'diabetes' in as legitimate, the algorithm recalibrates the probability until it comes down well below 95 per cent and the email arrives in your inbox.

These algorithms are built so that they begin to spot the other key words that mark out the junk diabetes email from the legitimate ones. The inclusion of the word 'cure' could well distinguish the duds. Machine learning means that the algorithm will go through every email that comes in, trying to find patterns and links until it ends up producing a bespoke algorithm suited to your own individual lifestyle.

This updating of probabilities is also how driverless cars work. It's really just a more sophisticated version of controlling the paddle in Breakout. Move the steering wheel right or left according to the pixel data the machine is currently receiving. Does my score go up or down as a consequence?

Biases and blind spots

There is something uncanny about the way the Netflix recommender algorithm was able to identify traits in films that we as humans would struggle to articulate. It certainly challenges Lovelace's view that a machine will always be limited by the insights of the person who programs it. Nowadays algorithms possess a skill that we don't have: they can assess enormous amounts of data and make sense of it.

This is an evolutionary failing of the human brain. It is why the brain is not very good at assessing probabilities. Probabilistic intuition requires understanding trends in experiments run over many trials. The trouble is that we don't get to experience that many instances of an experiment and so we can't build up the intuition. In some ways the human code has developed to compensate for our low rate of interaction with data. So it is possible that we will end up, thanks to machine learning, with codes that complement our own human code rather than replicating it.

Probabilities are key to much of machine learning. Many of the algorithms we considered in Chapter 4 were very deterministic in their implementation. A human understood how a problem worked and programmed a computer that then slavishly went through the hoops it had been programmed to jump through. This is like the Newtonian view of the world, in which the universe is controlled by mathematical equations and the task of the scientist is to uncover these rules and use them to predict the future.

The revelation of the physicists of the twentieth century is that the universe is not as deterministic as we originally thought. Quantum physics revealed that Nature plays with dice. Outcomes depend on probabilities, not clockwork. And this reign of probability is what has made algorithms so powerful. It may also be why those trained in physics appear to be better placed than us mathematicians to navigate our new algorithmic world. It's the rationalists versus the empiricists, and, unfortunately for me, the empiricists are coming out on top.

How did that machine learn to play the Atari game of Breakout without being told the rules of the game? All it had was knowledge of the pixels on the screen and a score and a paddle that could be moved left or right. The algorithm was programmed to calculate the effect on the score of moving left or right given the current state of the screen. The impact of a move could be several seconds down the line, so you have to calculate the delayed impact. This is quite tricky because it isn't always clear what causes a certain effect. This is one of the shortcomings of machine learning: it sometimes picks up correlation and believes it to be causation. Animals suffer from the same problem.

This is rather beautifully illustrated by an experiment that revealed pigeons to be superstitious. A number of pigeons were filmed in their cages and, at certain moments during the day, a food dispenser was moved into the cage. The door to the dispenser was on a delay so the pigeons, although excited by the arrival of the food dispenser, would have to wait to get the food. The fascinating discovery was that whatever random action the pigeon happened to be doing right before the door was released would be repeated the next day. The pigeon had seen that the door was closed. It spun round two times and then the door opened. It then (falsely) deduced that the spinning round was what had caused the door to open. Because it was determined to get the reward, the next time the feeder appeared it spun around twice for good measure.

Another classic example of bad learning that you will find chanted like a mantra in the machine-learning community took place inside the US military, when neural networks were used to train machines to pick out pictures with tanks in them. The team designing the algorithm fed it with pictures that they'd labelled as containing tanks or not, and by analysing the data the algorithm began to pick out the features that distinguish the two. After analysing several hundred labelled pictures the algorithm was then tested with a batch of photos that it hadn't seen before. The team were very excited that it performed with 100 per cent accuracy.

The algorithm was passed on to the army for use in the field. Within a short time the army sent the algorithm back and declared it useless. The research team were perplexed. When they analysed the images that the US army had used, they could see the algorithm was almost randomly making up its mind. Until someone spotted what it was doing, it seemed to be very good at detecting tanks if the photo was taken on a cloudy day.

Returning to their training data, they understood what had gone wrong. The research team had only had access to a tank for a few days. So they'd driven around taking lots of pictures of the tank in different camouflaged positions. What they hadn't spotted is that during those few days the weather had been overcast. After that, they returned and gathered images of the countryside without tanks. But this time there were clear skies. All the algorithm had picked up was an ability to distinguish between pictures with clouds and pictures with clear skies. A cloudy day detector was not going to be much good to the military. The lesson: the machine may be learning but you need to make sure it's learning the right thing.

This is becoming an increasingly important issue as algorithms trained on data begin to affect society. Mortgage applications, policing decisions and health advice are being increasingly produced by algorithms. But there is a lot of evidence now that they encode hidden biases. An MIT graduate student, Joy Buolamwini, was perturbed to find that the robotics software she was working with seemed to have a much harder time picking up her face than that of lighter-skinned colleagues. When she wore a white mask, the robot picked out her face immediately, but as soon as she removed the mask she disappeared.

The problem? The algorithm had been trained on lots of images of white faces. The data had not included many black faces. This bias in the data has led to a whole host of algorithms that are making unacceptable decisions: voice recognition software trained on male voices that doesn't recognise women's voices; image recognition software that classifies black people as gorillas; passport photo booths that tell Asians their photos are unacceptable because they have their eyes closed. In Silicon Valley, four out of five people hired in the tech industry are white males. This has led Buolamwini to set up the Algorithmic Justice League to fight bias in the data that algorithms are learning on.

The legal system is also facing challenges as people are being rejected for mortgages, jobs or state benefits because of an algorithm. These people justifiably want to know why they have been turned down. But given that these algorithms are creating decision trees based on their interaction with data that is hard to unravel, justifying these decisions is not easy.

Some have championed legal remedies, but they are devilishly hard to enforce. Article 22 of the General Data Protection Regulations introduced into EU law in May 2018 states that everyone shall have 'the right not to be subject to a decision based solely on automated processing' and the right to be given 'meaningful information about the logic involved' in any decision made by a computer. Good luck with that!

There have been calls for the industry to try to develop a meta-language that the algorithm can use to justify its choices, but until this is successfully done we may have to be more cautious about the impact of these algorithms in everyday life. Many algorithms are good at one particular thing but not so good at knowing what to make of irregularities. When something strange occurs, they just ignore it, while a human might have the ability to recognise the out-of-the-box scenario.

This brings us to the No-Free Lunch Theorem, which proves that there is no universal learning algorithm that can predict outcomes accurately under any scenario. The theorem proves that even if the learning algorithm is shown half the data, it is always possible to cook the rest of the unseen data so that it might generate a good prediction on the training data but be out of whack when it comes to the rest of the unseen data.

Data will never be enough on its own. It has to come paired with knowledge. It is here that the human code seems better adapted to coping with context and seeing the bigger picture – at least for now.

Machine versus machine

It is this power to change and adapt to new encounters that was exploited to make AlphaGo. The team at DeepMind built their algorithm with a period of supervised learning. This is like an adult helping a child learn the skills that the adult has already acquired. Humans make progress as a species because we accumulate knowledge which we pass on in a much more efficient manner than when it was first achieved. I am not expected to single-handedly rediscover all of mathematics to get to the frontier. Instead I spent a few years at university fast-tracking through centuries of mathematical discovery.

AlphaGo began by going through the same process. Humans have played millions of games of Go that have been recorded digitally online. This is an amazing resource for a computer to trawl through, gleaning which moves gave the winner an edge. Such a large database allowed the computer to build up an idea of the probability, given a particular board position, of success for each move on the board. The data is small when one considers all the possible paths each game might take, but it provides a good basis for playing – although its future opponent might not go down the path that the losing player did in the database, and that is why just using this data set wasn't going to be enough.

The second phase, known as reinforcement learning, is what gave the algorithm the edge in the long run. At this point the computer started to play itself, learning from each new game that it generated. As certain seemingly winning moves failed, the algorithm changed the probabilities that such a move would win the game. This reinforcement learning synthetically generates a huge swathe of new game data. And by playing itself the algorithm has a chance to probe its own weaknesses.

One of the dangers of this reinforcement learning is that it can be limited and self-reinforcing. Machine learning is a little bit like trying to climb to the top of Everest. If you don't know where you're going and you are blindfolded and are asked to climb to the highest peak, one strategy would be to keep taking small steps from the place you are standing to test if that step takes you higher.

This strategy will eventually get you to a point that is the highest in your local environment. Any move from that peak will take you back down again. But that is not to say that by going down you couldn't find that there is another far higher peak on the other side of the valley. This is the challenge of what are called local maxima, peaks that make you feel like you've got to the top but that are little more than tiny hillocks surrounded by towering mountains. What if AlphaGo maximised its game play to beat other players in this local maxima?

This appeared to have been the case when European Champion Fan Hui discovered a weakness in the way AlphaGo was playing some days prior to the event with Lee Sedol. But once the algorithm was introduced to this new game play it quickly learned how to revalue its moves to maximise its chances of winning again. The new player forced the algorithm to descend the hill and find a way to scale new heights.

DeepMind now has an even better algorithm that can thrash the original version of AlphaGo. This algorithm circumvented the need to be shown how humans play the game. Like the Atari algorithm, it was given the 19x19 pixels and a score and started to play, experimenting with moves. It exploited the power of reinforcement learning, the second stage in the building of AlphaGo. This is almost tabula rasa learning and even the team at DeepMind was shocked at how powerful the new algorithm was. It was no longer constrained by the way humans think and play.

Within three days of training, in which time it played 4.9 million games against itself, it was able to beat by 100 games to nil the version of AlphaGo that had defeated Lee Sedol. What took humans 3000 years to achieve, it did in three days. By day forty it was unbeatable. It was even able to learn in eight hours how to play chess and shogi, a Japanese version of chess, to such a level that it beat two of the best chess programs on the market. This frighteningly versatile algorithm goes by the name of AlphaZero.

David Silver, lead researcher on the project, explained the impact of this blank-slate learning in multiple domains:

> If you can achieve tabula-rasa learning you really have an agent which can be transplanted from the game of Go to any other domain. You untie yourself from the specifics of the domain you are in to an algorithm which is so general it can be applied anywhere. For us AlphaGo is not about going out to defeat humans but to discover what it means to do science and for a program to be able to learn for itself what knowledge is.

DeepMind's goal is to 'solve intelligence . . . and then solve everything else'. They believe they are well on the way. But how far can this technology go? Can it match the creativity of the best mathematician? Can it create art? Write music? Crack the human code?

7

PAINTING BY NUMBERS

The unpredictable and the predetermined unfold
together to make everything the way it is.
Tom Stoppard

A few years ago I wandered into the Serpentine Gallery one Saturday afternoon and was transfixed. I guess that sense of spiritual exhilaration is what we're after when we enter a gallery space. My companions were struggling to connect, but as I walked through the rooms I became obsessed with what I saw.

On show was Gerhard Richter's series *4900 Farben*. 'You've never heard of Gerhard Richter?' my wife asked incredulously as we took the train into town. 'He's only one of the most famous living artists on the planet.' She often despairs at my lack of visual knowledge, immersed as I am for most of the day in the abstract universe of mathematics. But Richter's project spoke directly to the world I inhabit.

The work consists of 196 paintings, each one a 5x5 grid of squares. Each square is meticulously painted in one of a possible twenty-five different carefully selected colours. That means there is a total of 4900 coloured squares, which is where the title of the exhibition comes from. Richter has different versions of the way the paintings can be displayed. In the Serpentine exhibition, he had chosen to show Version 2, in which the 196 paintings are hung in groups of four to make forty-nine canvases, each consisting of 100 = 10x10 coloured squares.

Staring at these pixellated canvases, your natural urge is to seek meaning in the collections of squares. I found myself focusing on the way three yellow squares aligned in one 10x10 block. We are programmed to search for patterns, to make sense of the chaotic world around us. It's what saved us from being eaten by wild animals hiding in the undergrowth. That line of yellow might be nothing, but then again it could be a lion. Many psychologists, like Jung, Rorschach and Matte Blanco, believed that the mind so hankers after meaning, pattern and symmetry that one can use such images to access the human psyche. Jung would get his patients to draw mandalas, while Rorschach used symmetrical inkblots to tap into the minds of his patients.

The desire to spot patterns is at the heart of what a mathematician does, and my brain was on high alert to decode what was going on. There were interesting pockets of squares that seemed to make meaningful shapes. As I drifted through the gallery from one grid to another, I started to wonder whether there might be another game going on beneath these images.

I counted the number of times I saw two squares of the same colour together in a grid, then the slightly rarer occurrence of a line of three or four squares of the same colour. Having gathered my data, I sat down and started calculating what you might expect to see if the pixels had all been chosen randomly. Randomness has a propensity to clump things together in unexpected ways. That is why, when you're waiting for a bus, you often experience a big gap followed by three red buses that come rolling along together. Despite the buses setting off according to a timetable, the impact of traffic soon creates randomness in the arrival of the buses.

I began to suspect that the three yellow squares I'd spotted were the result not of a deliberate choice but of a random process at work behind the creation of the pieces. If there are twenty-five colours to choose from and each one is chosen randomly, then it is possible to figure out how many rows should have two squares next to each other of the same colour. The way to calculate this is to consider the opposite. Suppose I were to pick red for the first square. The probability that the next square would be of a different colour is 24/25 since I've got to avoid picking red. The chances that the third square will be different from the colour I've just picked is again 24/25. So the probability of getting a row of ten colours which never has two squares of the same colour side by side is:

$$(24/25)^9 = 0.69$$

This means that in each 10x10 painting you should see three rows (and three columns) with two of the same colour side by side. Sure enough, the canvases did indeed match this prediction.

My calculations told me that you should find, in the 49x10 rows on display, six with three of the same colour in a column or row. Here I found that while the columns checked out, there were more rows than expected with three colours. But that's the point of randomness. It's not an exact science.

Later, after the show, I decided to investigate Richter's approach and discovered, sure enough, that the colours had been chosen at random. He had put squares of twenty-five colours into a bag, and had determined which colour to use next by drawing a square from the bag, and 196 different canvases were created in this way for the Serpentine Gallery show. The total number of possible canvases can be calculated as 25^{25} different paintings. This is a number with thirty-six digits! Laid end to end, the canvases would be 4.3×10^{31} kilometres long and would take us well outside the farthest visible reaches of space.

I think my wife regretted taking me to the Serpentine Gallery. For days after, I was obsessed by calculating the coincidences in the paintings. Not only that, given that the exhibition displayed just one way to put the canvases together, I began to fixate on how many other versions might be possible. Version 1 had all the canvases together in one huge 70x70 pixellated image. But how many other ways could you arrange them? The answer turned out to be related to an equation that had intrigued the great seventeenth-century mathematician Pierre de Fermat.

I couldn't resist sending my musings to the director of the Serpentine Gallery, Hans-Ulrich Obrist. Some time later, I received a letter from Richter asking if he could translate my thoughts into German and publish them alongside his images in a book he was producing. He said he had been unaware of quite how many mathematical equations were bubbling beneath the art he had made.

A similar process was used for Richter's design for the stained glass windows in Cologne Cathedral's transept. In the cathedral, however, there is an element of symmetry added, as Richter mirrors three of his randomly generated window designs to make up the six-window group. The symmetry isn't obvious, therefore, but it might tap into the brain's affinity for patterns rather like a Rorschach inkblot.

Richter had in some sense exploited a code to create his work. By giving up the decision about which colour would be used and letting the random fumbling around in the bag be responsible, Richter was no longer in control of what the result would be. There is an interesting tension here between a framework created by the artist but the execution carried out without the artist any longer being in control.

This use of chance would be one of the principal strategies in some of the early attempts to build creative algorithms, code that would surprise the coder. The challenge is to find some way of passing the Lovelace Test. How can you create something new, surprising, of value, which nonetheless goes beyond simply what the writer of the code put in at the beginning? The idea of mixing a deterministic algorithm with a dash of randomness as Richter had done was a potential way out of the Lovelace dilemma.

What is art?

But why would anyone want to use computers to create art? What is the motivation? Isn't art meant to be an outpouring of the human code? Why get a computer to artificially generate art? Is it commercial? Are the creators trying to make money by pressing 'print' and running off endless new pieces of art? Or is this meant to be a new tool to extend our own creativity? Why do we as humans create art? Why is Richter's work regarded as art while a book of Dulux colour samples is not? Do we even know what this thing we call art really is? Where did it all begin?

Although the human lineage emerges in Africa six million years ago, we see evidence of creativity only when the first tools materialise. Stones crafted to create a cutting tool began to appear 2.6 million years ago, but this moment of innovation does not seem to have sparked a great creative surge. The human drive to create art seems to emerge 100,000 years ago. Archaeological finds in the Blombos Cave in South Africa have revealed what archaeologists believe are paint-making kits. It's not clear what they used the paint for: painting their bodies? Painting designs on leather or other objects? Painting even on the walls? Nothing survives in these caves in South Africa, which were not ideal for the long-term preservation of rock art.

But other caves across the world that are deeper underground have preserved some of the earliest images created by humans. Images of hands appear on walls in a striking number of these caves. Research has shown that caves in Maros on the Indonesian island of Sulawesi have images painted by humans that date back 40,000 years. The artist is believed to have blown red ochre from the mouth using the hand as a stencil. When the hand was removed its outline remained.

It is an existential statement. As Jacob Bronowski expressed it in his famous TV series *The Ascent of Man*: 'The print of the hand says, "This is my mark. This is man."'

In addition to hands we find human figures, and pictures of wild hoofed animals that are found only on the island. An image of a pig has been shown to be at least 35,400 years old and is the oldest figurative depiction in the world. Scientists are able to date these images by dating the calcite crusts that had grown on top of the images. Because the crusts formed after the paintings were made, the material gives a minimum age for the underlying art. Something happened 40,000 years ago which unleashed a period of sustained innovation in the human species.

But humans might have been beaten to the first example of cave art by Neanderthals. Images of hands in caves in Spain had been regarded as dating from the period when *Homo sapiens* moved from Africa to Europe, 45,000 years ago, resulting in the European Neanderthals being wiped out as species some 5000 years later. But recent dating of the calcite crusts on some of the images in the caves in Spain dates the art to more than 65,000 years ago. Humans weren't in Europe. This is art created by another species. But both of these are beaten by designs carved on shells found on Java which date back as far as 500,000 years ago and are the work of *Homo erectus*, ancestor to both humans and Neanderthals. We thought art was uniquely human. But it appears we must share the discovery of art with Neanderthals and *Homo erectus*.

Some would argue that we shouldn't call this art. And yet it seems clear that this represents an important moment in evolution when a species started making marks with an intention that probably goes beyond mere utility. Experiments to re-create some of the carvings made in bone dating back 40,000 years reveal the staggering amount of labour that was expended on creating these pieces. Such extravagance on the part of a tribe hunting and surviving shows the value the carving must have had that they allowed the carver to skip their obligations. We will never know what the intention behind these creations truly was. The marks could have been made on a shell as a gift to impress a mate or to denote ownership, but it still represents an act that would evolve into our own species' passion for artistic expression.

The question of what actually constitutes art is one that has occupied humanity for centuries. Plato's definition of art in *The Republic* is very dismissive. Art is the representation of a physical object which is itself the representation of the abstract ideal object. For Plato art depends on and is inferior to the physical object it is representing, which in turn depends on and is inferior to the pure form. With this definition, art cannot yield knowledge and truth but leads to illusion.

Kant defines art as 'a kind of representation that is purposive in itself and, though without an end, nevertheless promotes the cultivation of the mental powers for sociable communication'. Tolstoy picks up on this idea of communication declaring art 'a means of union among men, joining them together in the same feelings, and indispensable for the life and progress toward well-being of individuals and of humanity'. From the caves of Altamira to the Serpentine Gallery, art has the potential to bind the individual into a group, revealing how our personal human code resonates with another's.

For Wittgenstein art is part of the language games that are central to his philosophy of language. They are all attempts to access the inaccessible: the mind of the other. If and when we can create a mind in a machine, then its art will be a fascinating way to penetrate what it feels like to be a machine. But we are still a long way from creating conscious code.

Art is ultimately an expression of human free will, and until computers have their own version of this, art created by a computer will always be traceable back to a human desire to create.

Even if the program is sparked into action by certain trigger words that it sees on Twitter, this can't be interpreted as the algorithm suddenly feeling like it must express a reaction. That has been programmed into the algorithm by the coder. The desire to create still lies in the mind of the human, even though it might not know when that action will be executed.

And yet modern views of art challenge whether it represents anything at all. Is it more about politics and power and money? Those who say it is art define art. If Hans-Ulrich Obrist decides to show a collection of work at the Serpentine Gallery in London, then his powerful position in the art world means that many will engage with the pieces in a way that without the meta-data of the curator's seal of approval might mean the pieces are missed.

Much modern art is no longer about the appreciation of an aesthetic and skill by the likes of Rembrandt or Leonardo, but rather the interesting message and perspective that the artist is revealing about our relationship to our world. Duchamp places a men's urinal in an exhibition space and the context makes a functional object into a statement about what is art. John Cage gets an audience to listen to 4 minutes and 33 seconds of silence and suddenly we begin to question what music is. We begin to listen to the sounds that creep in from the outside and appreciate them in a different way. Robert Barry's piece, pencilled in fine block letters on the wall, that simply reads: 'All the things I know but of which I am not at the moment thinking – 1:36 PM; June 15, 1969', challenges the viewer to negotiate the idea of absence and ambiguity. Even Richter's *4900 Farben* is really not about an aesthetic or his skill at painting squares. It is a political statement challenging our ideas of intention and chance.

So does computer art represent a similar political challenge? If you laugh at a joke, what difference does it make if subsequently you are told that the joke was created by an algorithm? The fact that you laughed is good enough. But why not other emotional responses? If you cry when you see a piece of art and then are told that the art is computer-generated, I suspect most people would feel cheated or duped or manipulated. But then it begins to question whether we are truly connecting with another human mind or just exploring untapped reaches of our mind. And yet this is the challenge of another's consciousness. All we have to go on is the external output of the mind, since we can never be truly inside another mind.

As Andy Warhol declared: 'If you want to know all about Andy Warhol, just look at the surface: of my paintings and films and me, and there I am. There's nothing behind it.'

But for many using computers in their art this is simply about a new tool. We never regarded a camera as being creative but rather as allowing a new creativity in humans. The school of computer art is experimenting in the same way, exploring whether the restrictions and possibilities take us in a new direction.

Creative critters

Given that we are going to explore creativity beyond the human realm, it seems worth pausing to consider whether there are any other species that have emerged through the evolutionary tree with a level of creativity to match our own.

In the mid-1950s Desmond Morris, a zoologist, gave a chimpanzee at London Zoo a pencil and a piece of paper and the chimp drew a line over and over again. Congo, as the chimp was known, soon graduated on to paintbrushes and canvases, and in 2005 three of his creations sold at auction for £14,400. That same auction saw a work by Andy Warhol go unsold. Did this make Congo an artist? Or would he, for that to be the case, have to have knowledge of what he was doing? The drive to create was primarily coming from Morris rather than Congo, so this should really be recognised as a disguised form of human creativity.

Some in the zoo community believe that giving tools to animals in captivity can relieve their stress and help avert the repetitive behaviour that animals in zoos so often resort to. Others have criticised zoos for cashing in on the products of animal creativity by selling canvases by elephants in the zoo shop or auctioning the painted handprints of lemurs on eBay. Zoo animals are a strange group to study, as their environment is so distorted. Can we find examples of animal creativity in the wild?

The Vogelkop gardener bowerbird builds elaborate towers out of grass that it decorates in a manner which seems to involve distinct choices. These constructions serve a purpose: to woo a female mate. They suggest the mastery of certain skills that are valuable to a mating couple, but the towers themselves go way beyond whatever proficiency may be needed to make a nest. So is the Vogelkop gardener bowerbird creative, or does the utilitarian nature of his endeavour throw his accomplishment into question?

Birds sing to communicate. But at some stage this skill developed to the point where they were able to do way more than was strictly necessary. Excess, demonstrating the ability to be wasteful, is of course a signal of power in animals and humans. So pushing oneself to be extravagant in making a nest or singing a song is a way of signalling one's suitability as a mate.

Interesting questions of copyright ownership have been raised when animals have been given tools to create. David Slater, who left a camera in the Tangkoko nature reserve in Indonesia to see if he could get the resident macaques to take photographs, was overjoyed when he developed the film to discover that the macaques had taken the most extraordinary selfies. When the pictures found their way on to the internet, he decided to sue the users for breach of copyright. It took a while for his case to wend its way to trial, but in August 2014 the US courts surprised him by denying him ownership of the pictures on the grounds that an item created by a non-human cannot be copyrighted. Things got more bizarre the following year, when the People for the Ethical Treatment of Animals, or PETA, filed a counter-suit against Slater for breaching the macaque's copyright. This case was thrown out of court.

The judge in the second case contended that for the macaque known as Naruto who took the selfie 'There is no way to acquire or hold money. There is no loss as to reputation. There is not even any allegation that the copyright could have somehow benefited Naruto. What financial benefits apply to him? There's nothing.' PETA was told in no uncertain terms to stop monkeying around.

How might these test cases apply to works created by AI? Eran Kahana, an intellectual-property lawyer at Maslon LLP and a fellow at Stanford Law School, explains that the reason why IP laws exist is to 'prevent others from using it and enabling the owner to generate a benefit. An AI doesn't have any of those needs. AI is a tool to generate those kinds of content.' What if AI creates a piece of art in the style of a living artist? It's likely that the programmer might be sued for copyright infringement but it's a very grey area. Inspiration and imitation are central to the artistic process. Where is the line between your own creation and copying someone else's?

When a film studio hires many people to create a movie it is the studio that owns copyright. Maybe AI will have to be given the same legal status as a company. These may seem like rhetorical abstractions but they are actually important issues: why would anyone invest in creating a complex algorithm that can compose new music or create art if the output could be used by anyone without cost? In the UK there has been a move to credit 'the person by whom the arrangements necessary for the creation of the work are undertaken'. In America the Copyright Office has declared that it will 'register an original work of authorship, provided that the work was created by a human being'. But will these laws need to change as code becomes more sophisticated? This brings us back to Ada Lovelace's question of whether anything new can really be created that transcends the coder's input. Are coders our new artists?

Coding the visual world

One of the first examples of visuals made by code that could hang in a gallery was created by Georg Nees, working at Siemens in Germany in 1965. The language that allows a computer to turn code into art is mathematics, but Nees was not the first to experiment with the relationship between maths and the visible world. It was the French philosopher René Descartes who understood that numbers and pictures were intimately related. Descartes created a way to change the visual world into a world of numbers and vice versa. Called Cartesian geometry, by drawing two perpendicular axes on the page every point can be identified by two numbers. These two numbers describe how much you should move horizontally and vertically along the axis to arrive at the location of the point.

This is the principle of a GPS coordinate. If I want to locate the position of my college in Oxford on a map, then the two numbers (51.754762, −1.251530) tell me how far I should move north and then west from the starting point at (0,0), which is the point where the line of longitude through Greenwich meets the equator.

Since every point you mark on a page can be described in terms of numbers, this allows us to describe any geometric shape we might draw by the numbers that describe all the points that make up that shape. For example, if you mark all the points where the second coordinate is twice the first coordinate, then all these points make up a line at a steep lean across the page. The equation for this is $y = 2x$. You could also specify that the first coordinate should be between two values, say $1 < x < 2$. Then you will get a short line at a lean.

I like to think of Descartes's ideas as a bit like a dictionary translating from one language to another. But instead of translating between French and English, Descartes's dictionary allows you to move between the language of geometry and the language of numbers. A geometric point gets translated into the numbers that define the coordinates of that point. A curve gets translated into the equation that defines the coordinates of all the points on that curve.

Descartes's dictionary which turns geometry into numbers was a revolutionary moment in mathematics. Geometry had been a mainstay of mathematics ever since Euclid introduced its axiomatic approach to the interplay between lines, points, triangles and circles, but now mathematicians had a new tool to explore this geometric world. The exciting thing was that although the geometric side of the dictionary was limited by our three-dimensional universe, the numbers side of the dictionary could be taken into higher dimensions. Things that could not be physically constructed could be imagined in the mathematician's mind, a concept that allowed mathematicians at the end of the nineteenth century to create new shapes in four dimensions. It was the discovery of these new imaginary geometries that inspired Picasso to try to represent hyperspace on a two-dimensional canvas.

The potential to use equations to manipulate these numbers, especially in the age of the computer, led to some interesting and surprising outcomes as Nees began to explore with his machines at Siemens. He programmed his computer to start at one point on the canvas and draw twenty-three lines that join up to make a shape. Each line would begin where the last line ended. The lines would alternate between heading off horizontally or vertically. To program this geometric output Nees needed to write the code using the numbers side of Descartes's dictionary. There were two random elements he introduced into the equation: whether to head up or down, left or right, and how long the line should be. The twenty-third line would close the shape up by connecting the end of line 22 to the starting position.

The result was curiously interesting. Nees arranged 266 of these images in a 19x14 grid. Positioned in this way, they looked like the designs Le Corbusier would draw in his notebooks. Nees could have done this by hand, but the power and ease of the computer to generate new iterations at the touch of a button allowed him to experiment with different rules and to experience their effect on a more accelerated timescale. His work revealed the computer to be a new tool in the artist's toolbox.

The random element Nees had introduced into the program meant that it could produce images he was not in control of and could not predict. This did not mean that the computer was being creative. Creativity is about conscious or subconscious choices, not random behaviour. Yet the constraints he had introduced, combined with randomness, had led to the creation of something that has enough tension to hold the eye.

One could argue that anything that doesn't have randomness programmed into it, that is deterministic, must still really be the creation of the programmer, regardless of the surprise the programmer might get at the outcome. But is this really fair? After all, there is some sense in which one might regard all human action as predetermined. There are real challenges as to whether humans really have the free will that we all believe we do.

The atoms in our bodies follow the equations of physics. Their position and movement at this moment in time determine what they will do in the future, bound on their course by the laws of Nature. That motion might be chaotic and unpredictable, but classical physics asserts that it is predetermined by the present. If atoms have no choices in what they do next, then we, who are made of atoms, have no choices. Our actions are predetermined by the code that controls the universe. If human action is predetermined, then are our creative acts any more our own than the computer's, which people claim belong to the programmer not to the computer?

Perhaps our only hope for agency in our actions is to appeal to the quantum world. Modern physics asserts that the only truly random thing happens at a quantum level. It is on the level of subatomic particles that there is some element of choice in the future evolution of the universe. What an electron is going to do next is random, based on how the quantum wave equation controlling its behaviour collapses. There is no way to know in advance where you will find the electron when you next look at it. Could the creativity of humans, which seems to involve choice, actually depend on the free will of the subatomic world? To make truly creative code, might one need to run the code on a quantum computer?

Fractals: nature's code

Nees believed that the closed loops he had created were just the beginning of the power of the computer to create visual art. In the ensuing decades computers allowed programmers to experiment, revealing extraordinary visual complexity in a simple equation. The discovery of the visual world of fractals, shapes with infinite complexity, would have been unthinkable without the power of the computer. As you zoom in on a fractal, rather than becoming simpler at small scale, it maintains its complexity. It is a shape that is in some sense scale-less because you cannot discern from a section at what scale of magnification you are.

The most iconic of these fractals is named after the mathematician who sparked the explosion of computer-generated images: the Mandelbrot set. Anyone who went clubbing in the 1980s would recognise this shape as the one that would be projected onto walls as DJs spun their psychedelic music. By infinitely zooming in on the image, the graphics created a sense of falling into some dreamlike world without ever touching the ground. These shapes could never have been discovered without the power of the computer. But is that art?

In his 'Fractal Art Manifesto', published in 1999, Kerry Mitchell tried to distinguish fractal art from something a machine was doing. The art, he argued, was in the programming, the choice of equation or algorithm, not the execution:

> Fractal Art is not . . . Computer(ized) Art, in the sense that the computer does all the work. The work is executed on a computer, but only at the direction of the artist. Turn a computer on and leave it alone for an hour. When you come back, no art will have been generated.

No claim is being made that the computer is being creative. One of the qualities that distinguishes fractal art from the computer art generated by Nees is that it is totally deterministic. The computer is making no choices that are not programmed in before it starts calculating. Why do computer fractal images, although new and surprising, still feel so anaemic and lifeless? Perhaps the answer lies in the fact that they do not form a bridge between two conscious worlds.

Computer-generated fractals have nonetheless made their creators big money, as fractals have proven to be a highly effective way to simulate the natural world. In his seminal book *The Fractal Geometry of Nature*, Benoit Mandelbrot explained how Nature uses fractal algorithms to make ferns, clouds, waves, mountains. It was reading this book that inspired Loren Carpenter, an engineer working at Boeing, to experiment with code to simulate natural worlds on the computer. Using the Boeing computers at night-time, he put together a two-minute animation of a fly-through of his computer-generated fractal landscape. He called the animation *Vol Libre*, meaning 'Free Flight'.

Although Carpenter was meant to be making these animations for Boeing's publicity department, his ultimate aim was to impress the bosses of Lucasfilm, creators of *Star Wars*. That was his dream: to create animations for the movies. He finally got his chance to show off his algorithmic animation at the annual SIGGRAPH conference held in 1980 for professional computer scientists, artists and filmmakers interested in computer graphics. As he ran his 16mm film, he could see in the front row of the audience the guys from Lucasfilm he was hoping to impress.

When the film came to an end, the audience erupted with applause. They hadn't seen anything so impressively natural created by an algorithm. Lucasfilm made him a job offer on the spot. When Steven Spielberg saw the effects that Carpenter was able to create with code, he was so impressed he declared: 'This is a great time to be alive.' Carpenter's colleague Ed Catmull concurred: 'We're going to be making entire films this way someday. We'll create whole worlds. We'll generate characters, monsters, aliens. Everything but the human actors will come out of computers.'

Carpenter and Catmull together with Alvy Ray Smith went on to found Pixar animations, which today employs as many mathematicians and computer scientists as it does artists and animators. The luscious jungle landscapes of a film like *Up* would once have taken artists months to produce. Today they can be created instantaneously at the click of an algorithm.

The power of fractals to create convincing landscapes using minimal code also makes the technology perfect for building gaming environments. It was Atari in 1982 that first recognised the potential of this technology to transform the gaming world. It invested a million dollars in the computer graphics department at Lucasfilm to convince them to help revolutionise the way games were made.

One of its first successes was a game released in 1984 called, appropriately enough, Rescue on Fractalus. The gaming environment is more forgiving than a motion picture, so the landscape could look less realistic and gamers would still be happy. The team were still rather frustrated with the jagged nature of the pixellation. But eventually they just accepted that this was as good as it was going to get on the Atari machines. They decided to embrace the jagged nature of the graphics, calling the aliens on Fractalus the 'jaggis'. But, as processing power on gaming machines advanced, so did the power of games to create more convincing worlds. To go from a static PacMan space to the almost movie-like rendering of games such as Uncharted is down to the power of algorithms.

Perhaps one of the most creative uses of algorithms in the gaming world was in the vast game called No Man's Sky, released in 2016. Developed for Sony's PlayStation 4, players roam round a universe visiting what seems like an endless supply of planets. Each planet is different, populated by its very own flora and fauna. Although they are perhaps not technically infinite in number, Sean Murray, who helped create the game, believes that if you were to visit a planet every second, our own real sun would have died before you visited them all.

So does Hello Games, the company that developed No Man's Sky, employ thousands of artists to create these individual planets? It turns out that there are just four coders, who are exploiting the power of algorithms to make these worlds. Each environment is unique and is created by the code when a player first visits the planet. Even the creators of the game don't know what the algorithm will produce before the planet is visited.

The algorithms being deployed at Pixar and PlayStation are tools for human creativity. Just as the camera didn't replace the portrait artists, computers are allowing animators to create their worlds in a new way. So long as computers are tools for human ingenuity and self-expression, they are no real threat to the artists. But what about computers that aim to create new art?

From AARON to 'The Painting Fool'

The artist Harold Cohen spent his life trying to create code that might be regarded as creative in its own right. Cohen began his career intending to be a conventional artist and seemed to be well on his way to achieving this goal when he represented Great Britain at the Venice Biennale in 1966, at the age of thirty-eight. Shortly after the show, he went to America and met his first computer thanks to an encounter with Jef Raskin at the University of California, San Diego. 'I had no idea it would have anything to do with art,' he said. 'I just got turned on by the programming aspect of the computer.' Raskin, who went on to create the Macintosh computer at Apple in the late seventies, turned out to be a great choice of teacher. (The name was chosen because McIntosh was his favourite variety of apple; the spelling had to be changed for legal reasons.)

Inspired by Raskin, Cohen went on to produce AARON, a code he wrote to make works of art. Cohen's code was of the top-down 'if . . . , then . . .' variety. By the time he died, it consisted of tens of thousands of lines. What was interesting to me was the language Cohen used to describe how the code chose what to create. He talked about AARON making decisions. But how had he programmed these decisions?

People involved in creating computer art tend to be reluctant to reveal the exact details of how their algorithms work. This subterfuge is partly driven by their goal of creating an algorithm that can't easily be reverse-engineered. It took me some digging in the code to reveal that 'making decisions' was Cohen's own code for the use of a random-number generator at the heart of the decision-making process. Like Nees, Cohen had tapped into the potential of randomness to create a sense of autonomy or agency in the machine.

Is randomness the same as creativity? Many artists find that a random occurrence can be a helpful spur to creation. In his *Treatise on Painting* Leonardo da Vinci described how a dirty cloth thrown at a blank canvas might serve as a catalyst for seeing something to inspire the next step. More recently Jackson Pollock allowed the swing of his bucket to determine his compositions. Composers have found that chance has sometimes helped them head in a new and unexpected direction in their musical composition.

But randomness has its limitations. There is no choice being made about why this configuration is more interesting than any another. Ultimately it will be a human decision to discard some of the output as less interesting than other parts. Randomness is of course crucial when it comes to giving the program the illusion of agency, but it is not enough. The 'on' button is still in the hands of humans. At what point will algorithmic activity take over and human involvement disappear? Our fingerprints will always be there, but our contribution may at some point be considered to be much like the DNA we inherit from our parents. Our parents are not creative through us, even if they are responsible for our creation.

But is randomness enough to shift responsibility from the programmer to the program?

Cohen died in 2016, at the age of eighty-seven. AARON, however, continues to paint. Has Cohen managed to extend his creative life by downloading his ideas into the program he created? Or has AARON become an autonomous creative artist now that Cohen is no longer on hand to partner his creation? If someone else presses the 'create button', who is the artist?

Cohen said he felt his bond with AARON was similar to the relationship between Renaissance painters and their studio assistants. Consider modern-day studios like those of Anish Kapoor or Damien Hirst, where many people are employed to execute their artistic vision. Kapoor has a big team in south London helping him, just as Michelangelo and Leonardo did.

Cohen was part of a whole movement of artists in the fifties and sixties who started exploring how emerging technology might unleash new creative ideas in the visual arts. The ICA in London held an influential exhibition in 1968 called *Cybernetic Serendipity*, which profiled the impact that the robotic movement was having in the art world. It included Nicolas Schöffer's *CYSP*

1, a spatial structure whose movements are controlled by an electronic brain created by the Philips Company. Jean Tinguely supplied two of the kinetic painting machines that he had developed called Métamatics. Gordon Pask created a system of five mobiles that interacted with each other based on the sound and light each emitted. The interactions were controlled by algorithms that Pask had written. The audience could also interact with the mobiles by using flashlights.

At the same time the Korean artist Nam June Paik was building his robot K-456, billed as history's first non-human action artist. It was created to perform impromptu street performances. As Paik recounted: 'I imagined it would meet people on the street and give them a split-second surprise, like a sudden show.' As technology has grown ever more sophisticated, so has the art exploiting that technology. But how far can these robots and algorithms go? Can they really become the creators rather than the creations?

Simon Colton has been working on a program to pick up the mantle from AARON. Here is what 'The Painting Fool', his creation, says about itself on its website:

> I'm The Painting Fool: a computer program, and an
> aspiring painter. The aim of this project is for me to be
> taken seriously – one day – as a creative artist in my own
> right. I have been built to exhibit behaviours that might
> be deemed as skilful, appreciative and imaginative.

Well, of course, these are the aspirations of Colton, its creator, rather than of the algorithm itself, but the aim is clear: for it to be considered a creative artist in its own right. Colton is not looking to use algorithms as a tool for human creativity so much as to move creativity into the machine. The Painting Fool is an ongoing and evolving algorithm that currently has over 200,000 lines of java code running its creations.

One of Colton's early projects was to create an algorithm that would produce portraits of people who visited the gallery. The results were then displayed on the walls of the gallery in an exhibition he called *You Can't Know My Mind*. The portraits needed to be more than just a photograph of the visitor taken by a digital camera. A portrait is a painting that captures something of the internal worlds of both the artist and the sitter. The artist in this case was an algorithm without an internal world, so Colton decided to algorithmically produce one. It needed to express (if not feel) some emotional state or mood.

Colton didn't want to resort to random-number generators to choose a mood, as that seemed meaningless. And yet he needed a certain element of unpredictability.

He decided to get his algorithm to read the large numbers of articles in the *Guardian* that day to set its emotional state. Certainly my morning perusal of the newspaper can lift or dash my mood. Reading about Arsenal's 4–2 defeat to Nottingham Forest in the FA Cup third round is likely to put me into a foul mood. My family knows to avoid me when that happens, while a preview of the final season of *Game of Thrones* might put me in a mood of excited expectation.

The programmers would not be able to predict the state of the algorithm, as they wouldn't know which article it would scan when it was prompted to paint. Yet there would be a rationale as to why the Painting Fool would choose to paint in a certain style.

When a visitor sat down for a portrait the algorithm would scan an article for words and phrases that might capture the mood of the piece. An article about a suicide bombing in Syria or Kabul would set the scene for a serious and dark portrait. Colton calls the choice 'accountably unpredictable'. The painting style isn't simply a random choice – the decision can be accounted for – but it is hard to predict.

Sometimes the Painting Fool would be so depressed by its

reading that it would send visitors away, declaring that it was not in the mood to paint. But before they left it would explain its decision, providing the key phrase from the article it had read that had sent it into such a negative tailspin. It would also stress: 'No random numbers were used in coming to this decision.'

This ability to articulate its decisions, Colton believes, is an important component of the dialogue between artist and viewer. In the exhibition each portrait comes with a commentary which seeks to articulate the internal world of the algorithm and to analyse how successful the algorithm thinks the output is in rendering its aims. These are two components Cohen said he missed in AARON.

I asked Colton if he believed that the creativity was coming from him and how much creativity he attributed to the algorithm. He very honestly gave the Painting Fool a 10 per cent stake in what was being produced. His aim is to change the balance of power over time.

He proposed a litmus test to this end: 'when The Painting Fool starts producing meaningful and thought-provoking artworks that other people like, but we – as authors of the software – do not like. In such circumstances, it will be difficult to argue that the software is merely an extension of ourselves.'

One of the problems Colton believes exists in mixing computer science and creative arts is that computer science thrives on an ethos of problem solving. Build an algorithm to beat the best player of Go. Create a program to search the internet for the most relevant websites. Match people up with their perfect partner. But creating art is not a problem-solving activity:

We don't solve the problem of writing a sonata, or painting a picture, or penning a poem. Rather, we keep in mind the whole picture throughout, and while we surely solve problems along the way, problem solving is not our goal.

In these other areas, the point of the exercise is to write software to think for us. In Computational Creativity research, however, the point of the exercise is to write software to make people think more. This helps in the argument against people who are worried about automation encroaching on intellectual life: in fact, in our version of an AI-enhanced future, our software might force us to think more rather than less.

The strategy of the team is to keep addressing the challenges offered by critics for why they think the output isn't creative, beating the critics finally into submission. As Colton puts it: 'It is our hope that one day people will have to admit that The Painting Fool is creative because they can no longer think of a good reason why it is not.'

AARON and the Painting Fool are both rather old school in their approach to creating art by machine. Their algorithms consist of thousands of lines of code written in the classic top-down mode of programming. But what new artistic creations might be unleashed by the new bottom-up style of programming? Could algorithms learn from the art of the past and push creativity to new horizons?

8

LEARNING FROM THE MASTERS

Art does not reproduce the visible; it makes visible.

Paul Klee

In 2006 a Mexican financier, David Martinez, paid $140,000,000 for a painting by Jackson Pollock known as *No. 5, 1948*. A number of incredulous critics questioned how flicking a load of paint around could command such prices. Surely this is something a kid could do!

It turns out Pollock's approach isn't quite so obvious as one might think. Pollock moved around a lot as he dripped the paint onto the canvas. He was often drunk and off balance at the best of times. The resulting image is a visual representation of the movement of his body as he interacted with paint and canvas. And yet that doesn't mean that it can't be simulated by a machine.

Mathematical analysis by Richard Taylor at the University of Oregon has revealed that Pollock's drip paintings are like a chaotic pendulum, where the pivot is allowed to move around instead of being fixed. And a chaotic pendulum is something I have studied and understand. I decided this was my chance to make millions by faking a Jackson Pollock. I rigged up a makeshift chaotic pendulum with a pot of paint on the end that would swing back and forth across the canvas I'd laid out on the floor, poured in some paint, and waited to see what would emerge.

The signature of chaos theory is a dynamical system that is incredibly sensitive to small changes, such that a seemingly imperceptible change in the starting position will result in a hugely different outcome. A conventional pendulum, as it swings back and forth, isn't chaotic. However, the pivot of my pendulum could be moved as the pendulum swung. This small shift suddenly caused it to behave chaotically. I had set up a system to mimic Pollock's physical movement as he painted. My chaotic pendulum was modelled on a contraption called 'The Pollockizer' that Taylor had designed to confirm his theory about Pollock's painting style.

The visual output that results from this chaotic paintpot is a fractal, an analogue version of the digital fractals exploited by Pixar and Sony to create their visual landscapes. The scale-less quality of a fractal is what makes Pollock's paintings so special. As you zoom in on a section, it becomes difficult to distinguish the zoomed-in-on section from the whole. Approaching the painting, you lose your sense of place in relation to the canvas and begin to mentally fall into the image.

Taylor's insight was a game changer. Many people over the years have tried to fake Pollocks by randomly flicking paint onto a canvas and selling them at auction as originals. But Pollock's fractal quality was something you could measure. With this insight mathematicians have been able to pick out the fake canvases 93 per cent of the time. However, I felt confident that the output of my chaotic contraption would pass the fractal test.

Our brains have evolved to perceive and navigate the natural world. Since ferns and branches and clouds and many other natural phenomena are fractals, our brains feel at home when they see these shapes. This is probably why Pollock's fractals are so appealing to the human mind. They are abstract representations of Nature. Recent research has confirmed that when participants are scanned in an fMRI scanner while looking at fractal images that are close to those we see in Nature, their parahippocampus region seems to activate. This is a region that is involved in regulating emotions and, interestingly, is often activated when we listen to music.

The recognition that similar parts of the brain fire whether we are looking at a Pollock or a fern or listening to music gets at one of the fundamental reasons why humans started to create art in the first place, and suggests why creativity is such an important and mysterious part of the human code. EEGs and fMRI scanners have given us a chance to penetrate the workings of the brain, but until then we were limited by our own senses and imagination. Pollock's paintings are portals into the way he sees the world around him. They come loaded with an implicit question: how do you see the world?

When I put my 'Pollock' up for sale on eBay, I was a little disappointed. I waited for a few hours, then a few days, finally a few weeks, but I got no bids! Locally the paint on the canvas looks like a Pollock but the problem is it has no structure. The chaotic pendulum produced drip fractals but was incapable of creating that overall impression of something more that Pollock was able to convey. This seems to be a fundamental limitation of many of the codes attempting to make art: they can capture detail at a local level but lack the ability to piece these bits together into a canvas that is satisfying on a larger scale.

Pollock's approach may appear mechanical, but he threw himself into every one of his paintings. 'It doesn't matter how the paint is put on,' he wrote in describing his method, 'as long as something is said. Painting is self-discovery. Every good artist paints what he is.'

Resurrecting Rembrandt

When Nees displayed his computer-generated art in the Stuttgart Academy of Fine Art back in 1965, the resident artists challenged him: 'Very fine and interesting, indeed. But here is my question. You seem to be convinced that this is only the beginning of things to come, and those things will be reaching way beyond what your machine is already now capable of doing. So tell me: will you be able to raise your computer to the point where it can simulate my personal way of painting?'

'Sure, I will be able to do this.' Nees replied. 'Under one condition, however: you must first explicitly tell me how you paint.'

Most artists are unable to explain how they create their art. This means that the process can't simply be coded up. The output is the consequence of many subconscious instincts and decisions. But could machine learning bypass the need for conscious expression by picking up patterns and rules that we are unable to detect? To test this proposition, I decided to investigate whether an algorithm could squeeze from beyond the grave just one more painting by one of the great artists of all time.

Rembrandt van Rijn was sought out for the skill with which he captured the emotional state of his subjects in his portraits, and his reputation has only grown over time. Many artists have viewed him as a paragon of their field, and despaired at ever reaching his level of skill and expressive mastery. As Van Gogh remarked: 'Rembrandt goes so deep into the mysterious that he says things for which there are no words in any language. It is with justice that they call Rembrandt – magician [sic] – that's no easy occupation.' He painted countless portraits of Dutch guild members and grandees as well as landscapes and religious commissions, but what most obsessed him were self-portraits, which he returned to again and again until his death, creating intimate autobiographical studies animated by a probing sincerity.

Was Rembrandt's considerable output sufficient for an algorithm to be able to learn how to create a new portrait that would be recognisably his? The internet contains millions of images of cats, but Shakespeare wrote thirty-seven plays and Beethoven nine symphonies. Will creative genius be protected from machine learning by a lack of data? Data scientists at Microsoft and Delft University of Technology were of the view that there was enough data for an algorithm to learn how to paint like Rembrandt. Ron Augustus from Microsoft, who worked on the project, believed the old master himself would approve of their project: 'We are using technology and data like Rembrandt uses his paints and brushes to create something new.'

The team studied 346 paintings in total, creating 150 gigabytes of digitally rendered graphics to analyse. The data gathering included detecting things like the gender, age and head direction of Rembrandt's subjects, as well as a more geometric analysis of various key points in the faces. After a careful analysis of Rembrandt's portraits, the team settled on a subject they felt typified a figure he might have drawn next: a 30- to 40-year-old Caucasian male with facial hair wearing dark clothes, a collar and a hat, facing to the right. It could easily have been a woman as there was almost a 50:50 split between the sexes, but the male portraits had more analysable details. You didn't really need a complicated data analysis to get to this point. Where machine learning came into its own was realising the portrait in paint.

The team used algorithms to explore Rembrandt's approach to painting eyes, noses and mouths. His use of light is one of the distinctive features of his paintings. He tended to create a concentrated light source on one area of the subject, almost like a spotlight. This has the effect of throwing some parts of the features into sharp focus while making other areas blurry.

The algorithm did not seek to fuse or create an average of all the features. As Francis Galton discovered in 1877, when he tried to construct a prototypical image of a convict by averaging photographs of real convicts, the result produces something far removed from the original. By layering the negatives on top of each other and exposing the resulting image Galton was rather shocked to see the array of distorted and ugly faces he had used transform into a handsome composite. It seems that when you smooth out the asymmetries, you end up with something quite attractive. The data scientists would have to devise a more clever plan if they were going to produce a painting that might be taken for a Rembrandt. Their algorithm would have to create new eyes, a new nose and a new mouth, as if it could see the world through Rembrandt's eyes.

Having created these features, they then investigated the proportions Rembrandt used to place these features on the faces he painted. This was something Leonardo had been fascinated by. His sketchbooks are full of measurements of the relative positions of different features in the face. Some believe he was tapping into mathematical ideas of the golden ratio to create the perfect face. Rembrandt was not so obsessed with the underlying geometry but nonetheless seemed to favour certain proportions.

The analysis was first conducted on flat images. But a painting isn't a 2D image. The paint on the canvas gives it a topography which contributes to the effect. For many artists this feature is as important as the composition. Think of how Van Gogh layered his oil paint, creating a sculpture as much as a painting. The textured quality of a painting is something that is frequently missed by those creating art via algorithms. The art is often rendered on a screen and is therefore limited to its 2D digital canvas. But what distinguishes artists from Goya to de Kooning is as much the way the paint is applied to the canvas as the image it produces. Certainly the way Rembrandt layers his paint is a key feature of his late output. But the team realised that modern 3D printers would give them a chance to analyse and sculpt the contours that are characteristic of Rembrandt's canvases. The final 3D printed painting consists of more than 148 million pixels, made from thirteen layers of paint-based UV ink.

Bas Korsten, one of the creative partners working on the project, admitted that while the idea was ingenious in its simplicity its execution was anything but. 'It was a journey of trial and error. We had plenty of ideas that were researched or tested, but discarded in the end.' The team had considered rigging up a robotic arm that would execute the final painting, but robotic arms currently have only nine degrees of freedom, while a human hand like Rembrandt's has twenty-seven different parts that can move independently. So that approach was abandoned.

The biggest challenge Korsten admits was keeping the idea behind 'The Next Rembrandt' alive. 'There were so many forces working against it. Time, budget, technology, critics. But, most of all, the overwhelming amount of data we needed to go through. Perseverance and not taking "no" for an answer are the only reasons why this project succeeded.'

After eighteen months of data crunching and 500 hours of rendering, the team finally felt ready to reveal to the world their attempt to resurrect Rembrandt. The painting was unveiled on 5 April 2016 in Amsterdam and immediately caught the public's imagination with over 10 million mentions on Twitter in the first few days of it going on display. The result is quite striking. There is no denying that it captures something of Rembrandt's style. If asked to name the artist, most people would probably put it in the Rembrandt school. But does it convey his magic? Not according to British art critic Jonathan Jones.

'What a horrible, tasteless, insensitive and soulless travesty of all that is creative in human nature,' he wrote with contemptuous disgust in the *Guardian*. 'What a vile product of our strange time when the best brains dedicate themselves to the stupidest "challenges", when technology is used for things it should never be used for and everybody feels obliged to applaud the heartless results because we so revere everything digital.'

Jones felt the project missed the entire point of Rembrandt's creative genius. It wasn't about style and surface effects, but the way Rembrandt was able to reveal his inner life, and in so doing unveil our own internal world. This was about two souls meeting. The AI painting totally failed to elicit what Jones calls the Rembrandt Shudder, that feeling one gets in front of every true Rembrandt masterpiece.

To his mind, there was only one way such a project could ever succeed: the AI would also have to experience plague, poverty, old age and all the other human experiences that make Rembrandt who he was, and his art what it is.

Is it fair to be so dismissive? Would he have reacted in the same way had he not been told ahead of time that a computer had produced the painting? The artist's process is often a black box. Algorithms have given us new tools to dig around inside the box and to find new traces of patterns. If we can replicate through code what an artist has done, then that code reveals something about the process of creation. Could that help us identify overlooked old masters or reattribute falsely catalogued works?

There has been much debate over the decades about who exactly painted *Tobit and Anna* in the Willem van der Vorm Collection in the Netherlands. It certainly has many of the characteristics of a late Rembrandt: concentrated light, a rough painting surface, parts that are very sketchy, together with others that are in sharp focus. It even has Rembrandt's signature at the bottom, but many believed that this had been added later and was fake. For decades it was not classified as a Rembrandt and was attributed to one of his pupils. This all changed in 2010, when a Rembrandt expert, Ernst van de Wetering, brought the powers of modern science to bear on the canvas.

Thanks to infrared scans and X-ray analysis we can now see things hiding beneath the surface, such as the first attempts the artist made on the way to the final painting as the work evolved.

The X-ray photographs revealed that initially the painting had included another window that was subsequently painted over. According to van de Wetering, Rembrandt continually played around with light in this way, trying out different ways to illuminate the figures. Microscopic chemical analysis can also reveal that the signature had to have been made while the painting was still wet. The combination of van de Wetering's years of experience and deep knowledge of Rembrandt's style, plus the support of these new scientific techniques, led him to change his mind about the attribution. The museum displaying the painting was very happy to hear that it had another Rembrandt in its collection, but there are still some critics who, despite the scientific support, still doubt the provenance of the work.

So what did van de Wetering think about this new computer-generated Rembrandt? He had hated the idea when it was first proposed. When he finally came face to face with the result he immediately started on a critique of the painting's brushwork, homing in on subtle inconsistencies and noting that the brushwork was the one Rembrandt was using in 1652, while the rest of the portrait was more in the style of work produced in the year 1632. The team was reasonably relieved that it was at this level of detail that their project was found wanting.

For Microsoft, the motivation for the Rembrandt project was most likely less artistic than commercial. To convincingly fake a Rembrandt demonstrates how good your code is. AlphaGo's triumph against Lee Sedol was similarly not so much about discovering new and more creative ways to play the game of Go as it was to provide great publicity for DeepMind's AI credentials. Is that a problem? Should creativity be free of commercial considerations? Van Gogh sold two paintings in his lifetime (although he did exchange other canvases for food and painting supplies from fellow artists). Perhaps he hoped to make a modest living, but money doesn't appear to have been much of a drive for his creativity. And yet there is evidence that dangling money in front of someone can stimulate (at least at a low level) their creative output.

In 2007 an American team of psychologists invited 115 students to read a short story about popcorn popping in a pan. The students were then asked to provide a title for the story. Half were told: 'We will be judging the creativity of your titles against the titles of all the other students who have participated in this research in the past. If your titles are judged to be better than 80 per cent of the past participants in this study, you will have done an excellent job.' The other half were told the same thing and given the prospect of a ten-dollar reward for their creativity. Sure enough, the financial incentive led to a far more creative output, including such gems as 'PANdemonium' or 'A-pop-calypse Now'.

Is feedback from others, whatever form it may take, an impetus for creation? Don't we continue to create and originate to keep our fellow humans engaged and interested in us? This is an aspect that the new AI is beginning to incorporate. In machine learning feedback is often used to move the algorithm towards a better result. Take DeepMind's algorithms for playing Atari games. Rewarding risk taking (by programming it to seek a high score) led the algorithm to crack levels that an unincentivised algorithm had missed.

Competitive creativity

Creating a new Rembrandt is fairly pointless beyond proving that it can be done. But could genuinely new and exciting art emerge from code? Ahmed Elgammal of Rutgers University wondered whether making artistic creation into a competitive game might spur computers into new and more interesting artistic territory. His idea was to create an algorithm whose job was to disrupt known styles of art, and a second tasked with identifying the output of the first as either not recognisably art or insufficiently original. It is a classic example of a General Adversarial Network, a concept first introduced by Ian Goodfellow at Google Brain. Each algorithm would learn and change based on the feedback from the other algorithm. By the end of the game, Elgammal hoped to produce an algorithm that would be recognised on the international stage for its creativity.

There is some evidence that this adversarial model is applicable to the way the human code channels creativity. This was suggested by the curious case of Tommy McHugh. In 2001 Tommy had a stroke. Before the stroke he had been happily leading his life as a builder in Liverpool. He was married and living in a small house in Birkenhead and had had no interest in art beyond the tattoos he'd decided to get while in prison. But after the stroke something strange happened. Tommy suddenly had an urge to create. He started writing poetry, but also bought paints and brushes and began to cover the walls of his house with artworks. The trouble was, he couldn't control this urge. He became a hostage to his drive to paint images on every wall in his home.

Stepping into the house is like entering a kitsch version of the Sistine Chapel. Everything is covered in pictures. Tommy's wife could not take the explosion of creativity and left him to it. Tommy couldn't stop. He just kept covering old paintings with new.

'Five times I've painted the whole house: floor, ceilings, carpets . . .' he told me. 'I only sleep through exhaustion. If I was allowed, the outside of this house would be painted and so would the trees and the pavements.'

Are the paintings any good? Not really. But why did Tommy suddenly have this urge to paint following the stroke? He tried to describe to me what was happening inside his head when this creative urge took hold: 'I kept on visualising a lightning flash shooting over to one side of the brain and hitting this one cell which unlocked a Mount Etna of bubbles, each little Fairy Liquid bubble in my imagination contained billions of other bubbles. And when they popped all this creativity exploded.'

Research by neuroscientists has discovered that, like the algorithms driving the Generative Adversarial Networks at Google Brain, our own brains have two competing systems at play. One is an exhibitionist urge to make things. To create. To express. The other system is an inhibitor, the critical alter ego that casts doubt on our ideas, that questions and criticises them. We need a very careful balance of both in order to venture into the new. A creative thought needs to be balanced with a feedback loop which evaluates the thought so that it can be refined and generated again.

It seems that Tommy's stroke knocked out the inhibitor side of his brain. There was nothing telling him to stop, that what he was creating might not be so great. All that was left was this explosive exhibitionist urge to create more and more crazy images and ideas.

The German artist Paul Klee expressed this tension in his *Pedagogical Sketchbook*: 'Already at the very beginning of the productive act, shortly after the initial motion to create, occurs the first counter-motion, the initial movement of receptivity. This means: the creator controls whether what he has produced so far is good.'

Tommy died in 2012 from cancer but had no regrets about what had happened to him: 'My two strokes have given me eleven years of a magnificent adventure that nobody could have expected.'

Elgammal's strategy was to write code to mimic this Generator/Discriminator dialogue that takes place generally subconsciously in an artist's mind. First he needed to build the Discriminator, an algorithmic art historian that would appraise the output. In collaboration with his colleague Babak Saleh, he began to train an algorithm so that it could take a painting it hadn't seen before and classify the style or painter responsible for it. WikiArt has probably the largest database of digitised images, with 81,449 paintings by 1119 different artists spanning 1500 years of history. Could an algorithm be created that could train itself on the content of WikiArt and take a painting at random and classify its style or artist? Elgammal used part of the available data as a training set and the remaining data to test how good the algorithm was. But what should he program his algorithm to look out for? What key distinguishing factors might help classify this massive database of art?

To use mathematics to identify an artist we need things to measure. The basic process is similar to the one behind the algorithm that drives Spotify or Netflix, but instead of personal taste you are looking for distinguishing characteristics. If you measure two different properties of the paintings in your data set, then each painting can be graphically represented as a point on a two-dimensional graph. So what can you measure that will result in you suddenly seeing Picasso's paintings clustered in one corner and Van Gogh's in another?

For example, measuring one feature (the amount of yellow used in a painting, for instance) might see paintings by Picasso (marked with an 'x') and Van Gogh (marked with an 'o') arranged on the scale like so:

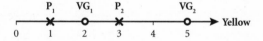

At the moment measuring this single feature doesn't help us distinguish between the painters. Sometimes Picasso uses a small amount of yellow, as in painting P_1, which scores a 1 on our scale. But at other times the yellow is more pronounced, as in painting P_2, which scores a 3. The two paintings VG_1 and VG_2 by Van Gogh plotted here also vary in the amount of yellow featured. Measuring yellow doesn't help us.

What if we pick another feature to measure (for example, the amount of blue in the paintings)? This time we'll plot the same paintings on a vertical axis now measuring this new feature.

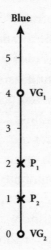

Again, blue doesn't really help us. There isn't a clear divide that puts paintings by Picasso on one side and Van Gogh on the other. But look what happens when we combine the two measurements, plotting the paintings now in two-dimensional space. Picasso's

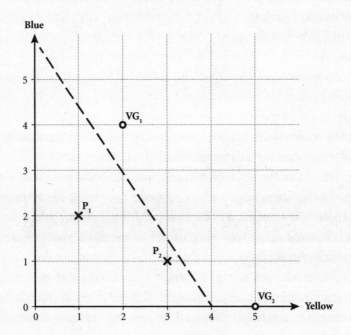

painting P_1 is located at position (1,2), while Van Gogh's painting VG_1 has position (2,4). But in this two-dimensional graph a line can be picked out which now partitions the paintings of each artist. We find by combining measurements of blue and yellow that Picasso's are in the lower half of the diagram while Van Gogh's appear in the upper half.

Having learned how to use these two features to distinguish a Picasso from a Van Gogh, when the algorithm is shown a new painting and told to identify whether it is by Van Gogh or Picasso, it will measure the two properties, plot the coordinates of the painting on the graph, and whichever side of the line the painting lands will give the algorithm the best bet as to the artist responsible for the painting.

In this simple example I've chosen the features of colour to distinguish the artists. But there are endless other features we could track. The power of machine learning is to explore the space of possible measurements and to pick out the right combination of features that help to differentiate between artists just as measuring yellow and blue did in our simple example. Two measurements will not be enough, so we need to find enough distinctive elements to tell one artist from another. Each new measurable feature increases the dimension of the space we are mapping paintings into and gives us a better chance of marking out artists and their style. By the end of the process we will be plotting paintings in a high-dimensional graph rather than in the two dimensions we saw in our simple example.

Finding the things to measure can be done in two different ways. As a programmer you can code up certain features that you think might help distinguish between artists: use of space, texture, form, shape, colour. But the more interesting feature of machine learning is its ability to engage in unsupervised learning and to find its own features to home in on. A human analysing the decision tree can sometimes find it hard to figure out what features the algorithm is focusing on to distinguish between paintings. The state-of-the-art in computer vision measures over 2000 different attributes in images that are now called classemes. These attributes were a good place to start to analyse the paintings the programmers had chosen to train their algorithm on.

In the quick sketch we cooked up on page 136 we saw how a two-dimensional space was sufficient to separate Picassos from Van Goghs. To get close to distinguishing styles across the true data set, the algorithm would have to plot paintings in 400-dimensional space, effectively taking 400 different sorts of measurement. The resulting algorithm, when tested on the unseen paintings, managed to identify the artists more than 50 per cent of the time but found it tricky to tell the difference between artists like Claude Monet and Camille Pissarro. Both are Impressionists who lived in the late nineteenth and early twentieth centuries. Interestingly, both artists attended the 'Académie Suisse' in Paris, where they became friends, a friendship that resulted in some noticeable interactions.

The Rutgers team decided to investigate whether their algorithm could identify moments in art history of extreme creativity, when something new appeared that hadn't been seen before. Could it identify paintings that had broken the mould and ushered in a new style of painting? Some artists are incrementally pushing the boundaries of an existing convention, while others come up with a completely new style. Could their algorithm identify the moment Cubism emerged on the scene, or Baroque art?

The algorithm had already plotted all the paintings as points in a high-dimensional graph. By adding the dimension of time to this graph and plotting when paintings were created, if the algorithm detects a huge shift in the position of the paintings in the high-dimensional space as it moves along this time dimension, does it correspond to a moment art historians would recognise as a creative revolution?

Take for example Picasso's *Les Demoiselles d'Avignon*, a painting that many acknowledge broke the mould. The initial reception when *Les Demoiselles* was first shown in Paris in 1916 was very hostile, as you would expect for a revolutionary change in aesthetic. A review published in *Le Cri de Paris* declared: 'The Cubists are not waiting for the war to end to recommence hostilities against good sense.' But it doesn't take long before the painting is being recognised as a turning point in art history. As the art critic at *The New York Times* wrote a few decades later: 'With one stroke, it challenged the art of the past and inexorably changed the art of our time.' The exciting thing is that the algorithm too was able to pick out a huge shift in the location of this painting compared to its contemporaries when viewed in this multidimensional graph, scoring it highly as a painting that was markedly different to anything that had gone before. Perhaps even the *New York Times*' art critic is about to be upstaged by an algorithm.

The Rutgers team's Discriminator algorithm is like an art historian who can judge whether paintings are part of an accepted existing style and recognise when they break new ground. Its counterpart, the Generator algorithm, is tasked with creating things that are new and different but will still be recognised and appreciated as art. To understand this tension between the new and the not too new, Elgammal steeped himself in the ideas of the psychologist and philosopher D. E. Berlyne, who argued that the psychophysical concept of 'arousal' was especially relevant to the study of aesthetic phenomena. Berlyne believed the most significant arousal-raising properties of aesthetics were novelty, unexpectedness, complexity, ambiguity, and the ability to puzzle or confound. The trick was to be new and surprising without drifting so far from expectation that arousal turned to aversion because the result was just too strange.

This is captured in something called the Wundt curve. If we are too habituated to the artwork around us, it leads to indifference and boredom. This is why artists never really stabilise in their work: what arouses the artist (and eventually the viewer) is

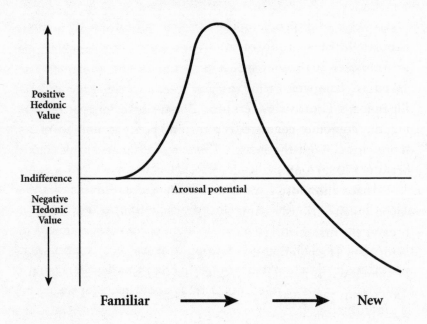

something distinct. The challenge is that the push to arousal or dissonance must not be so great that we hit the downslope of the Wundt curve. There is a maximum hedonic value that the artist is after.

Elgammal and his team programmed the Generator algorithm so that it was incentivised to create things which would try to hit that peak in the Wundt curve. The game was to maximise difference while trying not to drift too far from those styles that the art world has found acceptable. The Discriminator algorithm would be given the job of feeding back to the Generator algorithm whether it was too derivative or too wild to be considered art. Each judgement would alter the parameters of the Generator algorithm. This is machine learning in action: the algorithms change as they encounter more data, learning from the feedback. As the algorithms pinged information back and forth, the hope was that the Generator algorithm would be pushed to create new things that would fall in the sweet spot of the Wundt curve. Elgammal calls these Creative Adversarial Networks.

So what did people make of the output of these algorithms? When a group of art lovers were shown new works at Art Basel 2016, the flagship fair for contemporary art, and asked to compare them with new art works generated by Elgammal's Creative Adversarial Network, they reported finding the computer-generated art more inspiring, and identified more closely with the images. (To see the images: https://arxiv.org/abs/1706.07068.)

Perhaps the most significant signal that AI art is beginning to be taken seriously came in October 2018, when Christie's became the first auction house to sell a work of art created by an algorithm. The painting was produced by a Paris collective using Goodfellow's original idea of a General Adversarial Network rather than the Creative one developed by Elgammal. The Paris team trained their algorithm on 15,000 portraits dating from the fourteenth century through to the current day.

The result depicts a portrait of a man in a dark coat and white collar with unfinished facial features which gives the character a slightly unnerving quality. The portrait is strangely uncentred as if the sitter doesn't really want to be there. It is difficult to place the period of the painting, which combines an eighteenth-century style of portraiture with a very contemporary style of execution similar to the British artist Glenn Brown. The signature at the bottom is perhaps the most intriguing part of the whole painting. Instead of an artist's name we find a mathematical formula.

The portrait is one of a whole series produced by the algorithm which the Paris team have decided to put into a fictitious family tree depicting different generations of the Belamy family. The Christie's painting depicts Edmond Belamy, great-grandson of the Count de Belamy whose portrait was bought privately in February 2018 for $12,000. (In the Christie's auction, the portrait of the great-grandson went for a staggering $432,000.) The choice of family name pays homage to Goodfellow, who came up with the idea of these competing algorithms. Goodfellow translates loosely in French to 'Bel Ami'.

This idea of learning from what artists have done in the past and using that knowledge to push into the new is of course the process which most human artists go through. Current art can only be understood in light of our shared past. After all, this knowledge or frame of reference is what most viewers bring to their encounter with new art. No art on view at the Basel Art show is being experienced by someone who has never been exposed to the way Picasso and Munch have painted. Most creativity stems from this idea of perturbing the present to create a future that has some connection to the present but nonetheless breaks from it. It is an evolutionary model, and, intriguingly, this is what the algorithm picked up on.

You may feel this approach is horribly manipulative. To change art into a landscape of numbers only to find the points that will trigger maximal hedonic value sounds awful. Aren't great artists meant to express their inner angst? And yet there may just be a role for this alternative pathway to artistic creativity. These Adversarial Networks' algorithms can push us into new terrain that we recognise as art but have been too inhibited to explore. Computer code has the capacity to reveal untapped potential in the art created by the human code.

Seeing how an algorithm thinks

Art does many things, but for me where art is at its best is in providing a window into the way another mind works. And perhaps that is the true potential of art made by AI, because it might ultimately help humans to understand the hidden nature of the underlying computer code. If AI is due to take over from us humans, it might be a good idea to get some perspective on how AI views the world.

A team at Google have been using art created by AI to understand better some of the thought processes at work in the visual-recognition algorithms that they have been creating. As I explained in Chapter 5, the algorithms that have been developed to distinguish images of cats from bananas depend on a hierarchy of questions that the algorithm asks about the image. The algorithm effectively plays a game of twenty questions to identify what is in the picture.

The trouble is that, as the machine learns and changes, the programmer gradually begins to lose track of the features it is using to identifying bananas from cats. By looking at the raw code it is very difficult to reverse-engineer how the algorithm is working. There are millions of different questions that the algorithm can ask about an image, and it is tricky to see how and why these questions have been chosen in preference to others. To try to get a feel for how the algorithm was working, the team at Google had the clever idea of turning the program on its head. They gave the algorithm a random pixellated image and asked it to dial up or enhance the features it thought would trigger the recognition of an identifiable feature. They hoped that the result would reveal what the algorithm was looking for. They called this inverted algorithm DeepDream.

For me the images that DeepDream is producing are perhaps the most meaningful form of AI art that I've seen on my journey. Instead of trying to reproduce another Rembrandt or to compete with modern artists at the Basel Art show, these images are letting us see something of how visual-recognition algorithms view the world. It may not be very aesthetically important, but for me it is what art is all about: trying to understand the world through another set of eyes and to connect with a different way of seeing.

The DeepDream algorithm exploits the way a human can look at an image and suddenly see a face in their toast or an animal in the clouds when there is nothing there. The human brain has evolved to be extremely sensitive to images of animals because that is key to its survival. But this means sometimes we see animals where there are none. The visual-recognition algorithms work in a similar way. They are looking for patterns and interpreting them. They have learned to detect patterns in a compressed version of evolution, having been trained on thousands of images; their survival is dependent on correctly identifying them. Machine learning is basically a form of digital evolution. So what are the algorithms seeing in the digital undergrowth?

The results, the Google team discovered, were quite striking. Starfish and ants began to appear out of nowhere. It seems that within the algorithm was the power not just to recognise images but also to generate them. But this wasn't just a fun game. It offered fascinating insights into how the algorithm had learned. Images of dumbbells would always have an arm attached to the dumbbell. It was clear that the algorithm had learned about dumbbells from images of people lifting weights. So it hadn't understood that these things weren't an extension of human anatomy and could stand on their own.

Rather than feed the algorithm random pixels, you could give it actual images and ask it to enhance the features it detected or invite it to play the game we've all played of staring up at the clouds: what can you see hidden in those puffy shapes? The algorithm was able to pick out features that seemed to correspond to a dog or a fish or perhaps a hybrid animal. The novel that would become the cult film *Blade Runner* was called *Do Androids Dream of Electric Sheep?* Using these algorithms, we can now find out! In one image produced by the algorithm, sheep did indeed begin to appear in the sky.

More and more decisions are going to be taken out of human hands and placed in the digital hands of the algorithms we are making. The trouble is that the machine-learning algorithms that are appearing lead to decision trees that it is very hard for a human to unpick. This is one of the limitations of this new sort of programming. Ultimately we are not really sure why the algorithm makes the decision it does. How can we be sure that it isn't a mistake rather than an extremely insightful suggestion? The Go commentators were not sure which side of the divide to put AlphaGo's move 37 in game 2 until they eventually saw that it won the game. But increasingly these algorithms are doing more than playing games. They are making decisions that affect our lives. So any tools that help us to understand how and why these algorithms are making the decisions they do will be essential as we head into an increasingly automated future.

In the case of computer vision algorithms, the art they can produce is giving us some inkling as to how they are working. Sometimes the features that an algorithm is detecting and choosing between are things that we recognise, but at other times it seems hard to name what it is an algorithm is distinguishing in the image. The art is giving us an insight into the level of abstraction that an algorithm is working on at particular layers of the decision tree. We are penetrating what might be deemed the deep unconscious of that algorithm. The Google programmers called the process 'inceptionism' and considered the images to be like the dreams of the algorithm, hence the name 'DeepDream'. Certainly the images that their algorithm is generating have a crazy psychedelic feel to them, as if the algorithm were tripping on acid. By applying the algorithm over and over on its own outputs and zooming in after each iteration, the programmers could generate an endless stream of new impressions.

I don't think anyone would rank the product of DeepDream as good art (whatever that is). As the columnist Alex Rayner, who first wrote about these images, commented: 'they look like dormroom mandalas, or the kind of digital psychedelia you might expect to find on the cover of a Terrence McKenna book'. Not things you'll find at Frieze in London or at Art Basel. But it still represents an important new way of understanding something of the internal world of the algorithm as it classifies images.

The algorithm is the art

Are these new tools pushing the visual arts into interesting new territory? I decided I needed to make a trip back to the Serpentine Gallery to talk to Hans-Ulrich Obrist and hear his thoughts on the role of AI in the art world. But before heading up to his office I decided to have a peek at the art that was currently on show.

As I entered the gallery I was confronted by BOB, an artificial life form created with code by Ian Cheng. In fact there are six BOBs. Each started out with the same code, but the evolution of these life forms is affected by its interactions with visitors. By the time I made it to the exhibition, the six BOBs had gone off in very different directions. As a father to two genetically identical twin girls whose characters are very different, I know how a small change in the environment can have a large effect on the outcome of identical code.

Just as with Richter's *4900 Farben,* I felt compelled to unravel the code at the heart of BOB. But this is a different sort of code, one that is much harder to reverse-engineer. That may be why it succeeds in holding one's attention longer than one might expect. It is learning and evolving based on its interaction with the viewers who come to the gallery.

BOB picks up on the emotional state of the visitor via inter-actions with a smartphone. Cheng was intrigued by questions of authorship and origination. He wanted to know: 'How could art be authored in its meaning but also live beyond the author and mutate itself?' The answer was to create a system and allow its content to evolve and change based on interaction which he would not control. BOB's interactions with visitors mean that Cheng is left behind at some point as the code is informed by new parameters coming from its encounters.

Often we respond to code we don't understand by assigning it some sort of agency. When we didn't understand earthquakes or volcanoes we created gods that were responsible for these elusive forces. The algorithm at the heart of BOB stimulates the same response in the viewer in a phenomenon the philosopher Daniel Dennett refers to as the intentional stance.

As Hans-Ulrich told me:

> Usually the visitors' book at the gallery is full of
> complaints about the gallery being too hot or 'Why
> aren't there more chairs?' Or comments about how they
> like or don't like Grayson Perry. But we were getting
> instead comments like: 'Why doesn't BOB like me?
> I feel sorry for BOB. BOB ignored me. BOB is so cute.'
> It was extraordinary.

One night BOB appeared to have taken on a life of its own. Hans-Ulrich told me he had been travelling abroad a week before when he got a phone call from the security team at the gallery. At 3 a.m. the Serpentine had suddenly been flooded with light. Not a fire. Rather it appears that BOB had decided to wake up, despite the fact he'd initially been programmed to wake at 10 a.m. and to run until 6 p.m., when the gallery shut down. Our inability to understand why BOB woke up in the middle of the night makes us feel he has agency. It is this inability to understand how algorithms work that is fuelling the movies and stories of algorithmic apocalypse.

The open-ended nature of artwork that is continually evolving and never repeating, Hans-Ulrich believes, is something new for the art world. Most artworks that are hung in the gallery are static, frozen, physical objects that don't change over time or, at least, in the case of video art, have a beginning and an end. Any film in the gallery in the past would have to be looped and would ultimately become boring after you've seen it twenty times. The use of AI breaks that need to recycle material.

The code behind BOB shares something in common with the analogue code behind Jackson Pollock's drip paintings. It is based on chaotic deterministic equations that are influenced by the environment, so that the viewer can perturb the output. The chaos allows for unpredictability. Code that exploits the mathematics of chaos can claim to meet the criteria of novelty and surprise demanded by the word 'creative'. It remains deterministic, but chaotic processes are probably the best we can hope for if we intend to break the connection between coder and creator.

Jonathan Jones gave BOB one star in his *Guardian* review. 'They are just clever lab models. There is no soul here . . . art is always human, or nothing at all. Cheng forgets this, and his work is a techno bore.' Although Jones is almost certainly right that there is no ghost in the machine, as we head into the future we will increasingly need to exploit the world of the gallery as a mediator in understanding perhaps when the first ghost might appear.

Hans-Ulrich thinks of art as one of society's best early-warning systems. Given the importance of the debate about the role AI is playing in society, it seemed urgent in Hans-Ulrich's mind for AI to take its place in the gallery. Much of today's use of algorithms is invisible and hidden. We don't understand how we are being manipulated. Using art to visualise the algorithm helps us interpret and navigate these algorithms more knowingly. The visual artist is a powerful mediator between the crowd and the code. The artificial intelligence that was on display was the art.

'The artists are the experts at making the invisible visible,' Hans-Ulrich told me. So will AI ever create great art rather than being the art? 'We can never exclude that a great work can be created by a machine. One should never say never. As it stands today there hasn't been a great artwork created by a machine.' But he was cautious about the future: 'When the Go players said a machine is never going to beat us, Demis proved them wrong. I'm a curator but I'd never be arrogant and say a machine couldn't curate a better show . . .'

I could see his neurons beginning to fire. 'That could be a fun experiment to do one day . . . to do the Go experiment with curating . . . a dangerous experiment but an interesting one.'

9

THE ART OF
MATHEMATICS

*Sudden illumination is a manifest sign of
long, unconscious prior work.*

Henri Poincaré

I was thirteen when the idea of becoming a mathematician first took seed. The maths teacher at my comprehensive school took me aside after one lesson and recommended a few books he thought might interest me. I didn't really know at that stage what being a mathematician entailed, but one of those books revealed that it was much more than simple calculations. Called *A Mathematician's Apology*, the book was written by the Cambridge mathematician G. H. Hardy.

It was a revelation. Hardy wanted to communicate what it meant to do mathematics:

A mathematician, like a painter or a poet, is a maker of patterns. If his patterns are more permanent than theirs, it is because they are made with *ideas*. The mathematician's patterns, like the painter's or the poet's must be *beautiful*; the ideas like the colours or the words, must fit together in a harmonious way. Beauty is the first test: there is no permanent place in the world for ugly mathematics.

I'd never imagined mathematics to be a creative subject, but as I read Hardy's little book it seemed that aesthetic sensibilities were as important as the logical correctness of the ideas.

I wasn't much of a painter or a poet so why did my teacher think mathematics would be for me? When I got the chance to ask him many years later why he'd singled me out, he replied: 'I could see you responding to abstract thinking. I knew you'd enjoy painting with ideas.' It was a perfectly judged intervention that picked up on my desire for a subject that blended a creative mindset with a wish for absolute logic and certainty.

For years I've believed that the creative side of mathematics protected it from being automated by a computer. But now algorithms are painting portraits like Rembrandt and creating art works that rival human-generated painting on show at the Basel Art fair. Will they soon be able to re-create the mathematics of Riemann or compete with the papers published in the *Journal of the American Mathematical Society*? Should I start looking for another job?

Hardy spoke about mathematics like a game. He liked to use the analogy of chess, but ever since a computer could play chess better than humans, playing Go had always been my shield against those trying to quickly dismiss what I do as something a computer could do so much faster. Mathematics is about intuition, logical moves into the unknown that feel right even if I'm not sure quite why I have that feeling. But when DeepMind's algorithm discovered how to do something with a very similar flavour, it triggered an existential crisis.

If these algorithms can play Go, the mathematician's game, can they play the real game: could they prove theorems? One of my crowning pinnacles as a mathematician was getting a theorem published in the *Annals of Mathematics*. It is the journal in which Andrew Wiles published his proof of Fermat's Last Theorem. It is the mathematician's *Nature*. So how long would it be before we might expect to see a paper in the *Annals of Mathematics* authored by an algorithm?

In order to play a game it's important to understand the rules. What am I challenging a computer to do? I'm not sitting at my desk doing huge calculations. If that had been the case, computers would have put me out of a job years ago. So what is it exactly that a mathematician does?

The mathematical game of proof

If you read a news story about mathematics, it will invariably be about the fact that a mathematician has 'proved' some great outstanding conjecture. In 1995 newspapers ran breathless headlines about Wiles's proof of Fermat's Last Theorem. In 2006 the maverick Russian mathematician Grigori Perelman proved the Poincaré Conjecture, earning him the right to claim the million-dollar reward that had been placed on its head. There are still six more so-called Millennium Prize Problems that offer a challenge to prove thorny hunches mathematicians have about their subject.

The idea of proof is central to what mathematicians do. A proof is a logical argument that starts from a set of axioms, a list of self-evident truths about numbers and geometry. By analysing the consequences of these axioms one can start to piece together new statements that must be true about numbers and geometry. These new discoveries can then form the basis of new proofs, which in turn will invite us to discover yet more logical consequences of the axioms. This is how mathematics grows: like a living organism whose structure extends out from a previously existing form.

This is why people have compared mathematical proof to playing games like chess or Go. The axioms are the starting positions of the pieces on the board and the rules of logical deduction are the parameters determining how each piece can move. A proof is a sequence of moves played one after the other. In chess, given the number of possible moves at each stage, there are thousands of different positions the pieces can assume on the board. For example, after just four moves (two by white, two by black) there are already 71,852 different ways that the pieces can be arranged on the board. There are generally several different ways to reach that position. The tree of possible moves in Go grows even faster.

If I were to place the pieces randomly down on the board, you might ask, is it possible to reach this position from the starting position? In other words, is this a legitimate arrangement of pieces in a game of chess or Go? This is similar to the idea of a conjecture in mathematics. Fermat's Last Theorem, for example, was the conjecture that the equation $x^n + y^n = z^n$ can have no whole number solutions x, y and z when $n>2$. The challenge that mathematicians were faced with was proving whether this was a logical consequence of the way numbers work. Fermat placed the pieces on the board and said he believed you could get to this endpoint. Wiles and all the mathematicians who contributed to his work demonstrated a sequence of moves that ended with the arrangement Fermat had guessed was possible.

Part of the art of being a mathematician is picking out these targets. Many mathematicians believe that asking the right question is more important than providing the answer. Spotting what might be true about numbers requires a very sensitive mathematical nose. This is often where the most creative and difficult to pin down skill of the mathematician comes into play. It requires a lifetime immersed in this world to have that intuition about a possible new truth. It is often a feeling or hunch. You don't have to have an explanation for why this is true. That will be the proof everyone will be chasing.

This is one of the reasons why computers have found it hard to do mathematics. The top-down algorithms of the past are more like a drunk person stumbling around in the dark. They might randomly arrive at an interesting location, but most of the time their meanderings are unfocused and worthless. But could the bottom-up style start to develop an intuition of interesting locations to head for, based on past journeys made by human mathematicians?

How do mathematicians build up a feel for what might be an interesting direction in which to head? You may have some examples in mind to back up your hunch, a build-up of evidence that there seems to be a pattern here that is too good to be a coincidence. But patterns based on data can quickly vanish. This is why coming up with a proof is so important. It can sometimes take years to uncover a pattern as a false lead. I myself made a conjecture about a pattern in my own work that it took ten years for a graduate student to reveal to me was a false intuition.

One of my favourite examples of the danger of data is a hunch the great nineteenth-century mathematician Carl Friedrich Gauss had about prime numbers. He'd come up with a beautiful formula to estimate how many primes there were in the numbers from 1 to any number N, but he believed the formula he had found would always overestimate the number of primes. All the numerical evidence pointed to his being right. If a computer were to be let loose on the problem, it would to this day produce data to support Gauss's hunch. Yet in 1914 J. E. Littlewood proved theoretically why the opposite must be true. It turns out Gauss's guess only underestimates the primes after you've counted through more numbers than there are atoms in the universe (and even that won't get you anywhere near the point where the conjecture breaks down).

That is the challenge of all these conjectures. We just don't know if they are true or if our intuition and the available data are leading us astray. That is why we obsessively try to build a sequence of mathematical moves to link the conjectured endgame to legitimate games established to date.

But what is it that has driven humans to want to find these proofs? Where has the human urge come from to create mathematics? Is this motivation to explore the mathematical terrain one we will need to program into algorithms that will challenge mathematicians at their own game? The origins of mathematics, of course, go back to human attempts to understand the environment we live in, to make predictions about what might happen next, to mould our environment to our advantage. Mathematics is an act of survival by the human species.

The origins of mathematics

Mathematicians are a bit of a misunderstood breed. Most people think that as a research mathematician I must be sitting in my office in Oxford doing long division to lots of decimal places or multiplying six-digit numbers together in my head. Far from being a super-calculator, something a computer is clearly much better and faster at doing, a mathematician, as G. H. Hardy first explained to me, is at heart a pattern searcher. Mathematics is the science of spotting and explaining patterns.

It's this ability to spot a pattern that gives humans an edge in negotiating the natural world, because it allows us to plan into the future. Humans have become very adept at spotting these patterns, because those who missed the pattern didn't survive. When people I meet declare (as alas so often happens): 'I don't have a brain for maths', I counter that in fact we all have evolved to have mathematical brains because our brains are good at spotting patterns. Sometimes they are too good, reading patterns into data where none exists, as many viewers did when confronted with Richter's random coloured squares at the Serpentine Gallery.

For me, some of the very first pattern recognition comes along with some of the very first art to be drawn. The cave paintings in Lascaux depict exquisite images of animals racing across the walls. The movement of a stampede of aurochs is amazingly captured in these frozen images. It is intriguing to ask why the artist felt compelled to represent these images underground. What role did they play?

Alongside these images are what I believe to be some of the earliest recorded mathematics. A cluster of dots is understood to represent the constellation of the Pleiades, which is highest in the sky in the Northern hemisphere in the summer. There is then a strange series of dots, thirteen in number, ending with a great picture of a stag with huge antlers above the thirteenth dot. Next you get twenty-six dots ending with a picture of a pregnant horse. What is this abstract sequence of dots depicting? One conjecture is that each dot represents a quarter of a moon. Thirteen quarters of the moon represent a quarter of a year. So perhaps these dots are depicting a season, telling the viewer that a season on from sighting the Pleiades high in the sky is a good time to hunt stag, because at that time they are rutting and vulnerable.

In order to relay this information, someone had to spot that a pattern of behaviour seemed to repeat itself each year and that the pattern of animal behaviour corresponded to the pattern of moon phases. The drive to recognise such patterns was clearly practically motivated. There is utility driving the discovery.

Here we see the first ingredient of mathematics: the concept of number. Being able to formulate an accurate sense of numbers has been crucial to the survival of many animals. It will inform whether you fight or fly in the face of a rival pack. Sophisticated experiments have been done on newly born chicks that reveal quite a complex number ability hardwired into the brain. The chicks were able to judge that five is more than two and less than eight.

But to give these numbers names and represent them by symbols is a uniquely human ability. Part of our mathematical development has involved finding clever ways to identify or name these numbers. The ancient Mayans represented numbers with dots. They depicted the number by writing something down with the same number of elements. But at some point this becomes inefficient because it's hard to distinguish five dots from six dots. So someone had the clever idea of putting a line through the four dots to indicate five dots, just like a prisoner counting down the days till their release on the wall of the jail.

The Romans used a system whereby as numbers got bigger they were given new names: X for ten; C for a hundred; M for a thousand. The Ancient Egyptians too used new hieroglyphs to indicate another zero on the end of a number: a heel bone for ten; a coil of rope for 100; a lotus plant for 1000. But this system quickly gets out of hand as we get into the millions or billions. You need new symbols for each new large number.

The Mayans, who were doing sophisticated astronomy, needed big numbers to keep track of large swathes of time. They came up with a clever way to overcome the Roman problem. Called the place value system, it is the system we use today to write down big numbers. In our decimal system the positions of the digits indicate that they correspond to different powers of 10. Take the number 123. Here we have 3 units, 2 lots of 10 and 1 lot of 100. There is nothing special about the choice of 10 beyond the fact that we can use our fingers to count up to 10. Indeed, the Mayans had symbols all the way up to 20, and the position of a digit counted powers of 20. So 123 in Mayan mathematics denoted 3 units, 2 lots of 20 and 1 lot of $20^2 = 400$, making a total of 443.

The Mayans were not the first to come up with this clever idea of using the position of a number to indicate that it is counting different powers of 10 (or 20 in the case of the Mayans). Four thousand years ago the Ancient Babylonians had come up with this idea of the place value system. Instead of counting up to 20 like the Mayans, or in decimal as we do today, the Babylonians had symbols all the way up to 59 and then they started a new column. The choice of 60 was influenced by the high divisibility of this number. It can be divided by 2, 3, 4, 5, 6, 10, 12, 15, 20 and 30. This makes it a very efficient choice for doing arithmetic.

Necessity, efficiency and utility were driving these mathematical choices. We see their repercussions today in the way we keep track of time: 60 minutes in the hour, 60 seconds in the minute. Napoleon tried to get the measurement authorities to track time using a decimal system, but fortunately that never caught on.

In the cuneiform tablets the Ancient Babylonians left behind we witness the first mathematical analysis of how these numbers relate to the world around us. More sophisticated mathematics was born soon after, in conjunction with the growth of the city states around the Euphrates. To build, to tax, to do commerce require mathematical tools. These tablets reveal that officials were tabulating, for example, the number of workers and days necessary for the building of a canal, to calculate the total expenses of wages of the workers. There is nothing particularly challenging or interesting about the mathematics being done at this stage, but it clearly got some scribes thinking about what else you could do with these numbers.

They started to discover clever tricks to help them with their calculations. We find tablets with all the square numbers written out. These tablets were aids for calculating how to multiply large numbers together, because someone had noticed the interesting relationship between multiplying numbers and adding their squares. Captured by the algebraic relationship

$$A \times B = ((A + B)^2 - (A - B)^2)/4$$

the scribe realised that you could use these tables of squares to work out $A \times B$. First add A and B and look up the square of the answer and then subtract from that the square of $A-B$. Then divide the answer by 4. What is so exciting here is that this is a very early example of an algorithm at work. Here is a method that reduces the job of multiplying two numbers A and B together to the simpler task of adding and subtracting the numbers followed by using the database of squares contained on the tablet of squares. It works whatever the numbers A and B are, provided the squares don't exceed those calculated on the tablet.

Although the Babylonians were tapping into an algebraic way of thinking about numbers, they were far from having the language to articulate what they were doing. The equation I have written down only became possible thousands of years later when the Arabic and Persian scholars in the House of Wisdom in ninth-century Iraq developed the language of algebra. The Ancient Babylonians did not start writing down why this method or algorithm always gave the right answer. It worked and that was good enough. The curiosity to come up with a way to explain why it always worked would come later. This is why the word 'algorithm' comes from the chief librarian and astronomer at the House of Wisdom, Al-Khwarizmi, who founded the subject of algebra even though the first algorithms can be found in Ancient Babylonia.

This mathematical relationship between numbers is again being driven by utility. It speeds up calculation. It bestows an advantage on any merchant or builder who spots the connection. But we begin to see creeping in problems and ways to solve them that, although outwardly seemingly practical, if you consider them more closely they appear more like fun puzzles to challenge fellow scribes than something a farmer might want to know. For example, the following problem certainly sounds very practical:

A farmer has a field whose area is 60 units squared. One side of the field is 7 units longer than the other. What is the length of the shortest side of the field?

But here's the thing . . . How do you know the area of the field without knowing the lengths of the sides? For me this feels much more like someone setting a cryptic crossword puzzle. I'm thinking of a word but I'm only going to give you a rather mixed-up description of the word. You've got to undo that to work out the word I'm thinking of. In the case of the scribe's problem about the farmer, the unknown length of the field we can call X. The longer side of the field is then of length X + 7. The area of the field is these two lengths multiplied together and the answer we know is 60. So we get an equation:

$$X \times (X + 7) = 60$$

Or:

$$X^2 + 7X - 60 = 0$$

For some that will send a shiver of recognition through them because it's an example of the quadratic equations that students at school have to learn to solve. You can blame the Babylonian scribe for the challenge, but also thank him for the method the Babylonians concocted for unravelling this cryptic equation to find out what X is.

But for me this is an important transitional moment in my subject. Why did someone think of the challenge? Why did someone feel compelled to bother finding a clever way to unravel the challenge and find the answer? Why do we still get students to learn this? Not because you need to know it, as challenges like this don't really come up in everyday life. It is possible that the farmer had previously calculated the area and written it down but then forgotten to record the side of the length of the field, but why would he know that the long side of the field was 7 units longer while not knowing the length of the short side? The whole thing is too contrived ever to have been a genuine practical problem. No . . . this is doing maths for the fun of it!

This is a brain that is enjoying the 'aha' moment and delighting in how to untangle the problem to get the answer. We now know that a shot of dopamine or adrenaline would have accompanied the realisation that the method works whatever the numbers involved. There are biology and chemistry at work driving this mathematical move. Would a computer have made such a move to do mathematics purely for the fun of it, given that it has no biology or chemistry?

True, one could argue for an evolutionary advantage bestowed on the person who can do this sort of mathematics. And indeed this is our best defence for why we still insist on teaching students in school how to solve a quadratic equation. A mind that can apply this sort of algorithm, that can chase through the logical steps required to get to the answer, that is happy with an abstract, analytical thought process, is a mind that is well equipped to cope with problem solving in real life.

Perhaps the chemistry behind the satisfaction we feel when we solve a mathematical puzzle is going to be key to distinguishing human creativity from machine creativity. A brain is very much like a computer in its construction. And it might be possible to simulate the brain by creating an abstract network of digital neurons, switching on and off with relation to the other neurons connected to it. But if we don't put chemistry and biology into our construction, will we miss giving the machine that satisfying 'aha' moment that the Babylonian scribe was after? Will it lack the motivation, the drive, to think creatively?

In Babylonian mathematics you still find a focus on particular arithmetic examples. Methods discovered were applied to solve these particular problems, but no explanation was given for why these methods always worked. That would have to wait a few millennia, until mathematics started to develop the idea of proof.

The origins of proof

The beginning of this game of mathematical proof goes back to the Ancient Greeks, who discovered the power of logical argument to access eternal truths about number and shape. Proof is really what mathematics is about. This is the holy grail that a mathematician is in search of to make his or her name. To earn the million-dollar prize you've got to prove that one of the seven conjectures is true. To win a Fields Medal you've got to come up with a proof that impresses your fellow mathematicians. And it is probably Euclid's *Elements* that represents the rule book that kicked off this great game.

Coming back to our chess analogy will help us explain the way the game of mathematical proof works. You have an opening set of statements called axioms which are a little bit like the way we lay out the pieces at the beginning of a chess game. Euclid's *Elements* begins with these axioms. They are a list of things about numbers and geometry that mathematicians regard as blindingly obvious. Things we can all accept as true. It's of course true that we might be wrong about the truth of these axioms, but in some sense that doesn't matter for the game we are going to play. We just take these axioms as truths. And when you look at the sort of things Euclid included, it's fair to say they seem pretty acceptable as fundamental truths.

Between any two points you can draw a line. If $A = B$ and $B = C$ then $A = C$. Given any straight-line segment, you can draw a circle with that straight line as the radius of the circle. $A + B = B + A$.

Now we know how to lay the pieces on the board, we need to learn next how to play the game. Just as the chess pieces are constrained by certain rules which determine how they can move, there are rules for logical deduction that allow us to write down new truths based on what we know to date. For example, the rule Modus Ponens asserts that if you have established that statement A must imply statement B, and you've also established that statement A is true, then you are allowed to deduce that statement B is true. The complementary rule of Modus Tollens asserts that if you've shown that statement A must imply statement B, but now you've also learned that statement B is false, then you can deduce that statement A is false.

This last rule is applied in Euclid's *Elements* to prove that the square root of 2 cannot be written as a fraction. If it can be written as a fraction then by playing the game of mathematical chess and chasing a set of logical moves, eventually we get to the conclusion that odd numbers are even. But we know that odd numbers are not even. Therefore by applying the rule of Modus Tollens we arrive at the conclusion that the square root of 2 cannot be written as a fraction.

For me, the mark of a well-constructed and satisfying game is one that is simple to set up and whose rules are simple to understand and implement, and yet the range of games you can play is extremely rich and varied. Noughts and crosses is simple to explain and play but very soon becomes rather dull, because you start repeating the same games you've already played. In chess and Go, on the other hand, so many different games can evolve from the starting position that people who dedicate their lives to playing these games never tire of playing another game.

One important distinction between playing games like chess or Go and playing the game of mathematical proof is that mathematicians don't have to reset the pieces every time they want to play. All the games that have been played before become the foundations, the starting point from which you can play your next game. In a way the previous generations of mathematicians have expanded the axioms from which you can start, or the moves you can play, because anything that has been established to date can be used in the new game that you are about to play.

It's striking that we give meaning to these symbols and words. A line is that thing we draw across the page. X is meant to represent a number that counts or measures something. So how would a computer know what we're talking about? The beauty of the game is that even though we are trying to capture how number and geometry work, we can view the whole game symbolically.

In fact any meaning we give to the symbols such that the axioms are true will give rise to a game that teases out properties of the objects we have substituted for the symbols. This means a computer can make deductions about the game without really having to know what the symbols mean.

Indeed, when the nineteenth-century mathematician David Hilbert lectured on geometry he stressed this point: 'One must be able to say at all times – instead of points, lines and planes – tables, chairs and beer mugs.' His point was that provided the things had the relationship expressed by the axioms, then the deductions would make as much sense for chairs and beer mugs as for geometric lines and planes. This allows the computer to follow rules and create mathematical deductions without really knowing what the rules are about. This will be relevant when we come later to the idea of the Chinese room experiment devised by John Searle. This thought experiment explores the idea of machine translation and tries to illustrate why following rules doesn't show intelligence or understanding.

Nevertheless, follow the rules of the mathematical game and you get mathematical theorems. But where did this urge to establish proof in mathematics come from? A little bit of experimenting will reveal that every number can be written as prime numbers multiplied together and there always seems to be only one way to break down the number. For example 105 is equal to the product of primes 3 x 5 x 7, and no other combination of primes multiplied together will give you 105. You could just make the observation and hope that it always works. More examples just go to confirm your faith in this discovery. Indeed, you might start to think that the evidence is overwhelming and after a while could even suggest adding it as an axiom.

But what if there is some really large number where suddenly there are two different ways to pull it apart? It's just that you've got to hit really large numbers before this becomes possible. I think this is a point where we can distinguish a quality of mathematics that marks it out as different from science. A scientist would have to rely on evidence and data gathering to convince other scientists to add this as a good theory of the way numbers work. But the existence of proof means that we can show this is a logical consequence of the way numbers work. We can prove that there won't be an exceptional number that breaks the theory. Mathematical proof will show you why there is only one way to write numbers as products of prime numbers. And that proof will allow the next person who plays the game to include this as a given about the way numbers work.

The Babylonians would have been happy with the observation about numbers decomposing into products of primes but wouldn't have felt compelled to come up with a watertight argument for why this must always be true. They had a more scientific approach to numbers and geometry. It was the Ancient Greeks who came up with a new game, who marked out mathematics as a subject that allowed us to establish truth.

So where did this urge to prove come from? It is quite possible this is a by-product of the evolution of society from the cities of Ancient Egypt and Babylon, where power is centralised, to the new cities emerging in Ancient Greece where democracy, a legal system and political argument are part of everyday life. It is in Greece that we see writers beginning to use logical argument to challenge received opinion and authority.

In the stories that appear during this period humankind is no longer happy to be pushed around by the Olympian gods and begins to dispute the terms on which the gods want to rule them. Socrates, for whom an unexamined life is not worth living, dedicates his writing to argue the difference between truth and received opinion. Sophocles has Antigone challenge the tyrannical rule of her uncle. Aristophanes satirises the absolute power wielded by politicians in his democratic comedies.

This challenging of authority, this move to democracy and a society based on a legal system, necessitates developing skills of logical argument. The growth of the *polis*, which gave citizens a role in their society, necessitated the development of new skills to be able to engage in debate. Indeed, the sophists would tour cities giving lessons to people in rhetoric. In *The Art of Rhetoric* Aristotle defines rhetoric as 'the faculty of observing in any given case the available means of persuading'. He crystallises what tools a citizen needs, which include *logos*: the skill of using logical argument and available facts to persuade the crowd.

The drive to come up with clever forms of mathematical proof is triggered by this shift in society. *Logos* gave you the power to persuade. And that is why this push to use logical argument to persuade your fellow citizen of your point of view goes hand in hand with a shift in mathematics. The tools of logical deduction turned out to be powerful enough to access eternal truths about the way numbers and geometry work. You could prove that every number could be uniquely written as a product of prime numbers. You could prove that prime numbers go on to infinity. You could prove that a triangle subtended on a diameter of a circle was right-angled.

Very often you would have a hunch about one of these eternal truths. This is your conjecture that would arise from playing around with numbers. If you added up all the odd numbers in sequence you always seemed to get a square number: $1 + 3 = 4$, $1 + 3 + 5 = 9$, $1 + 3 + 5 + 7 = 16$. But does that always work? The Greeks were not content with observing this interesting possibility of a connection between odd numbers and square numbers. They wanted to use their new tool of *logos* to prove that this had to be the case; that it was the logical consequences of the basic axioms governing how numbers work.

And so began the great journey that is mathematics. Euclid's *Elements* set the scene for 2000 years of mathematicians coming up with proofs explaining the strange and wonderful way numbers and geometry work. Fermat proved why if you raise a number to the power of a prime bigger than that number and then divide the result by the prime, the remainder will be the number you started with. Euler proved why when you raise e to the power of i times pi the answer is -1. Gauss proved that every number can be written as the sum of at most three triangular numbers (writing 'Eureka' next to his discovery). And eventually my colleague Andrew Wiles proved that Fermat was right in his hunch that the equations $x^n + y^n = z^n$ don't have solutions when n>2.

These breakthroughs are representative of what it is a mathematician does. A mathematician is not a master calculator but a constructor of proofs. So here is the challenge at the heart of this book: why can't a computer join the ranks of Fermat, Gauss and Wiles? A computer can clearly out-perform any human when it comes to calculation, but what about our ability to prove theorems? It is possible to translate a proof into a series of symbols and a rule set for why one set of symbols is allowed to follow another. As Hilbert explained, you don't have to know what the symbols mean to be able to construct mathematical proofs. So doesn't this seem like something perfectly conceived for a computer to engage in?

Every time a mathematician takes an established mathematical statement and plays an allowed logical step, the new sequence of symbols represents a freshly established mathematical statement. It's possible that it's already on the list of mathematical proven statements because we came on it via a different route. But this is nonetheless a way for a mathematician (or a computer) to start generating new theorems out of old ones. Isn't that the goal? Mathematics may not be about doing calculations, but on this score hasn't the computer already put the mathematician out of a job if I can just press go and it starts spewing out logical consequences of all known statements?

Here is where creativity comes into the picture. It is easy to make something new. The top-down style of programming will produce a machine that can crank out new mathematical theorems. The challenge is to create something of value. Where does that value come from? This is something that depends on the mind of the human creating and consuming the mathematics. How will an algorithm know what mathematics will cause that exciting rush of adrenaline that shakes you awake and spurs you on?

This is why the new bottom-up style of programming emerging from machine learning is so exciting and potentially threatening for a mathematician like me: because these algorithms that Hassabis and his colleagues are producing can learn from the human mathematics of the past to distinguish the thrilling theorems from the boring ones, and this in turn might guide a machine on its way to producing a new theorem of value that might surprise the mathematical community just as the gaming world was so shocked by AlphaGo.

10

THE MATHEMATICIAN'S TELESCOPE

Our writing tools participate in the writing of our thoughts.
Friedrich Nietzsche

For all my existential angst about the computer putting me out of the game, I must admit as a tool it has proved invaluable. I can be faced with combining a slew of equations into one single equation. If I tried to do this by hand, I would almost certainly make a mistake. It is a mechanical procedure that requires little thought: you just have to follow a set of rules. This is something my laptop will not bat an eyelid at, and I would trust its calculation each time over my own pen-and-paper attempts. But the role that a computer might play beyond simply manipulating equations has grown over the years.

Given the close bond between mathematics and algorithms, it perhaps isn't surprising that computers have been partners in proving deep theorems of mathematics for nearly half a century. In the 1970s a computer played a major role in settling the proof of a classic challenge called 'the Four-Colour Map Problem'. No matter how you redraw the boundaries of the countries in Europe, the problem theorised, you could colour them all using at most four colours in such a way that no two countries with a common border have the same colour. It is impossible to colour all maps with three colours, but four would always suffice.

There was a proof that five colours would suffice, but no one had been able to reduce this to four. Then, in 1976, two mathematicians, Kenneth Appel and Wolfgang Haken, announced that they had found a way to prove why you could always get away with four colours. There was an interesting twist to their proof: they demonstrated that although there are infinitely many possible maps you can draw, there is a way to show that they can all be reduced to an analysis of 1936 maps. But analysing this many maps by hand was going to be impossible – or, to be more accurate, impossible for a human. Appel and Haken managed to program a computer to go through the list of maps and check whether each one passed the four-colour test. It took over 1000 hours for the lumbering computer of the 1970s to run through all the maps.

The computer was not doing anything creative. It was doing dumb donkey work. But could you prove that there wasn't a bug in the program giving false results? This question of how much we can trust the results of a computer is one that forever dogs the field of AI. As we head into a future dominated by algorithms, ensuring that there are no undetected bugs in the code will increasingly be a challenge.

In 2006 the *Annals of Mathematics* published a computer-assisted proof of another classic problem in geometry: Kepler's Conjecture. Thomas Hales, the human behind the proof, had come up with a strategy to prove conclusively that the hexagonal stacking of oranges you see at the grocer's is the most efficient way to pack spheres. No other arrangement wastes less space. Again, like Appel and Haken, Hales had used a computer to run through a finite but huge case analysis. He announced the completion of the proof in 1998 and submitted the paper to the *Annals of Mathematics*, along with the code he'd used for the computer component of the proof.

Before a paper is accepted for publication, mathematicians demand that all of its steps be checked by referees, running the proof like a program in their brains to see if anything crashes. But here was a bit of the proof that the brain with its physical limitations was unable to navigate. The reviewers had to trust the power of the computer. Many were uneasy about this. It was like wanting to go from London to Sydney and being forced to put your trust for the first time in an aeroplane for part of the journey. Because of the role the computer had played, it took eight years for mathematicians to agree that the proof was correct with 99 per cent certainty.

For mathematical purists, that 1 per cent was anathema. Imagine proving you're related to Newton . . . except for one missing link in the family tree. The role of the computer in proving theorems was viewed by many in the field with deep suspicion. Not because they felt nervous that it would put them out of a job – in these early years the computer could only work at the behest of the mathematician who'd programmed it – but because they were concerned as to how anyone could ever tell if there was a bug buried deep in the program. How could we trust such a proof?

Mathematicians had been burned by such bugs before. In 1992 Oxford physicists used heuristics from string theory to make some predictions about the number of algebraic structures that could be identified in high-dimensional geometric spaces. Mathematicians were a little suspicious – how could physics tell them about such abstract structures? – and they felt justified in their doubts when a proof showed that the conjecture was false. However, it turned out that the proof involved a computer component that was based on a program that had a bug in it. It was the mathematicians, not the physicists, who had been wrong. The bug in the program had led them astray. A few years later mathematicians went on to prove that the physicists' conjecture was correct (this time without a computer).

Stories like this have fuelled mathematicians' fears that computers might lead us to build elaborate edifices on top of programs that are structurally unsound. But frankly a human has more chance of making a mistake than a computer. It may be heresy to admit it, but there are probably thousands of proofs with gaps or mistakes that have been missed. I should know: I've subsequently discovered that a couple of proofs I'd published had holes in them. The gaps were pluggable, but the referees and editors had missed them.

If a proof is important, scrutiny generally wheedles out any gaps or errors. That is why the Millennium Prizes are released within two years of publication: twenty-four months is regarded as enough time for a mistake to reveal itself. Take Andrew Wiles's first proof of Fermat's Last Theorem. Referees spotted a mistake before it ever made it to print. The miracle was that Wiles was able to repair the mistake with the help of his former student Richard Taylor. But how many incorrect proofs might be out there leading us to build our mathematical edifices on falsehoods?

Some new proofs are now so complex that mathematicians fear hard-to-pick-up errors will be missed. Take the Classification of Finite Simple Groups, a theorem close to my own research. This is a sort of periodic table of the symmetrical atoms from which all symmetrical objects can be built. It is called the 'Monster Theorem' because the proof is 10,000 pages long and spans 100 journal articles and involved hundreds of mathematicians. The list of atoms includes twenty-six strange exceptional shapes called sporadic simple groups. There's always been a sneaking suspicion that a twenty-seventh might be out there that the proof may have missed. Could a computer help us check such a complex proof?

Besides, if you get a computer to check through a proof and verify that each step is valid, aren't you just moving the goalposts? How do we know that the computer program doing the checking doesn't have a bug? You could get another computer to check that program for bugs, but where would this end? Science and mathematics have always been dogged by this dilemma. How can you be certain that your methods are leading you to true knowledge? Any attempt to prove that it is invariably depends on the methodology you are trying to show produces truth.

As Hume first pointed out, much of science relies on a process called induction: inferring a general law or principle from the observation of particular instances. Why is this a sound way of generating scientific truths? Principally because of induction! We can point to many cases where this inductive principle seems to lead to good scientific theories. This leads us to conclude (or induce) that induction is a good approach to doing science.

Coq: the proof checker

As more and more proofs started to appear that were dependent on computer programs, it was felt that some approach was needed to ensure that the conclusions of these programs could be trusted. In the past, mathematics generated by humans could be checked by humans. Now it would be necessary to create new programs to check the programs behind the proofs, as their calculations were too complex and long for humans to validate.

In the late 1980s two French mathematicians, Pierre Huet and Thierry Coquand, began to work on a project called Calculus of Constructions, or CoC. In France there is a tradition of naming research development tools after animals, so the system soon came to be called Coq, the French for a male chicken. It was also, rather conveniently, the first three letters of one of the developers' surnames. Coq was created to check proofs, and soon emerged as the program favoured by anyone interested in validating computer proofs.

Georges Gonthier, the principal researcher at Microsoft Research Cambridge, decided to put together a team to use Coq to check the proof for the Four-Colour Map Theorem, the first proof that had required a computer for it to be completed. By 2000 the Microsoft Research team had run through the computer code developed by Appel and Haken and validated the proof (based on your trusting Coq not to have its own bugs). Then they set Coq the task of also checking the human-generated part of the proof, which Appel and Haken had written themselves.

One of the challenges in checking a human proof is that it rarely gives all the steps. People do not write proofs like computer code. They write proofs for other people, using a code that only has to work on our hardware, the human brain. This means that when we write proofs we often skip tedious steps, knowing that those reading the proof will know how to fill them in. But a computer requires every step. It's the difference between writing a novel, where you don't need to account for every tedious action of your central protagonist, and instructing a new babysitter, where you have to spell out every single detail of the day including naps, potty breaks and every last item on the menu.

It took another five years before the computer was able to verify the human element of the proof. An interesting by-product of this process was that the researchers uncovered new and rather surprising nuggets of mathematics that had been overlooked in the first proof.

Why, though, should we trust Coq any more than the original computer proof? The answer, interestingly, is because of induction. As Coq validates more and more proofs that we are confident are correct, we grow ever more certain that it has no bugs. That is really the same principle we use to test the fundamental axioms of mathematics. The fact that every time we take two numbers A and B we get the same answer if we add $A + B$ or

$B + A$ has led us to accept it as an axiom that $A + B = B + A$. By relying on one computer program to check all the others, we gain more trust in its conclusion than we can ever have in a bespoke program created especially for the proof at hand.

Once his team had finished checking the Four-Colour Map Theorem, Gonthier announced a new challenge to his team: the Odd Order Theorem. This is one of the most important theorems guiding the study of symmetry. Its proof led to the Classification of Finite Simple Groups, a list of the basic building blocks from which all symmetrical objects can be built. One of the simplest building blocks in this periodic table of symmetry is the regular two-dimensional polygons with a prime number of sides, shapes like the triangle or the pentagon. But there are many more complex and exotic examples of symmetry, from the sixty rotations of an icosahedron to the symmetries of a strange snowflake in 196,883-dimensional space that has more symmetries than there are atoms that make up the Earth.

The Odd Order Theorem states that any symmetrical object with an odd number of symmetries will not require exotic symmetries for it to be built. It can be made out of the simple ingredients of a prime-sided polygon. It was an important theorem because it essentially sorted out half the objects you might consider. From then on you could assume that the objects you were hoping to identify had an even number of symmetries.

The proof was pretty daunting. It ran to 255 pages and occupied the whole volume of the *Pacific Journal of Math*. Before its publication, most proofs covered at most a few pages and could be mastered in a day. This one was so long and complex it was a challenge for any mathematician to master. Given its length, doubts must always exist as to some subtle error that might be embedded inside the pages of the proof.

Getting Coq to check the proof would thus not only demonstrate Coq's prowess: it would contribute to our confidence in the proof of one of the most complex theorems in mathematics. This was a worthy goal. But turning a human-generated proof into checkable code expands it even further. Gonthier's challenge was not going to be easy.

He recalled sheepishly:

> The reaction of the team the first time we had a meeting and I exposed my grand plan was that I had delusions of grandeur. But the real reason for having this project was to understand how to build all these theories, how to make them fit together, and to validate all of this by carrying out a proof that was clearly deemed to be out of reach at the time we started the project.

One of his programmers left the meeting and looked through the proof. He emailed Gonthier his reaction: 'Number of lines – 170,000. Number of definitions – 15,000. Number of theorems – 4,300. Fun – enormous!' It took a total of six years for the team at Microsoft Research Cambridge to work through the proof. Gonthier spoke of the elation he felt as the project came to a close. At last, after many sleepless nights, he could relax.

'Mathematics is one of the last great romantic disciplines,' he said, 'where basically one genius has to hold everything in his head and understand everything all at once.' But we are reaching capacity with our human bit of hardware. Gonthier hopes his work will kick-start a period of greater trust and sustained collaboration between human and machine.

The limits of our human hardware

There is a growing sense among young mathematicians that many regions of the mathematical landscape are becoming so dense and complex that you could spend all three years of your PhD just trying to understand the problem your research supervisor has set you. You can spend years navigating this terrain, mapping out your discoveries only to find no one has the head space to retrace your steps to understand or verify them.

There is not much reward to reworking someone else's discoveries. And yet journals depend on this process of peer review. Promotion and tenure rely on the validation that getting a paper published in the *Annals of Mathematics* or *Les Publications mathématiques de l'IHES* bestows. So increasingly there could be a place for a system like Coq to help verify the proof of a theorem submitted to a journal.

Some mathematicians feel we are at the end of an era. The sort of mathematics that the human brain can navigate must inevitably have limits. Frankly I find it extraordinary how much mathematics has been within the reach of the human mind.

Take the Classification of Finite Simple Groups, the building blocks of symmetry. That we humans were able to construct, using our minds, pencils and paper, a symmetrical object that can only be built by working in 196,883-dimensional space is extraordinary. The mathematicians who truly feel at home working with the Monster Symmetry Group are growing old. Like the masons of the medieval period, they have skills that will be lost once they die. There isn't much compulsion for those who follow to rework these Gothic masterpieces unless they provide a pathway to new wonders.

That hundreds of pages of journals spanning three centuries combine to prove that Fermat's equations have no solutions is a testament to the long game that the human mind can play. And yet, when you are working to prove a conjecture, there will always be that sneaking feeling that the proof might command a complexity that is beyond the physical limitations of the human brain. It is amazing what we can do, but, given that mathematics is infinite and we are finite, we can prove mathematically that mathematics is bigger than we will ever be.

I am working on a conjecture now that has had me entangled in its grip for fifteen years. Every time I try to piece together the insights I've had on parts of the problem, my brain returns an error message that it's hit capacity. I have come tantalisingly close but I just can't pull the pieces together. I've been here before and know that finding a new way to look at a wild beast of a problem can often bring it within the net my mind is casting. When generations of mathematicians have been working on proving something like the Riemann Hypothesis, our greatest unsolved problem about prime numbers, without success, it's inevitable that someone would begin to wonder if, although the statement of the conjecture is simple enough, the proof might be beyond the limits of the human brain.

After spending years battling vainly with the Riemann Hypothesis, G. H. Hardy wryly pointed out: 'Every fool can ask questions about prime numbers that the wisest man cannot answer.' Kurt Gödel, the Austrian logician, has proven that mathematics has true statements for which there are no proofs. At some level this is a shocking revelation. Do we need to add new axioms to capture these unprovable truths? Gödel warned in 1951 that modern mathematics was likely to slip further and further from our grasp:

> One is faced with an infinite series of axioms which can
> be extended further and further, without any end being
> visible. It is true that in the mathematics of today the
> higher levels of this hierarchy are practically never used
> . . . it is not altogether unlikely that this character of
> present-day mathematics may have something to do with
> its inability to prove certain fundamental theorems, such
> as, for example, Riemann's Hypothesis.

Given that we may be reaching full capacity as humans, some mathematicians are beginning to acknowledge that if we want to push further, we'll need the machines. We could get to the top of Everest with little more than a tank of oxygen, but we never would have reached the moon without the union of human and machine.

One of those who believes the days of the lone mathematician working with pencil and paper are coming to an end is Doron Zeilberger, an Israeli mathematician who has teamed up with a computer to write papers since the 1980s and insists on including his machine, Shalosh B. Ekhad, as co-author on any paper in which he has used his computer. Shalosh B. Ekhad is Hebrew for 3B1, the name of the AT&T machine from which his current computer emerged. Zeilberger believes that the resistance to partnering with machines is due to what he calls 'human-centric bigotry', which, like other forms of bigotry, has held back progress.

Most mathematicians believe that their aspirations are more complex than those of computers: they hope to produce not just truths but an understanding of what lies behind those truths. If a computer verifies the truth of a statement without providing that understanding, they feel cheated.

'We aim to get understanding in mathematics,' said Michael Atiyah, who won the Fields Medal, the mathematical equivalent of the Nobel Prize. 'If we have to rely on an unintelligible computer proof, it's not satisfactory.' Efim Zelmanov, another Fields Medal winner, agrees: 'A proof is what is considered to be a proof by all mathematicians, so I'm pessimistic about machine-generated proofs.' Certainly, we don't accept a proof if only one mathematician can understand it. So does Zelmanov have a point? If only the machine that generated it can understand a proof, can we really trust it?

Doron Zeilberger appreciates where this sentiment comes from but ultimately dismisses it. 'I also get satisfaction from understanding everything in a proof from beginning to end,' he concedes. 'But, on the other hand, that's life. Life is complicated.'

He believes that if a human mind can understand a proof, then it must be pretty trivial:

> Most of the things done by humans will be done easily
> by computers in twenty or thirty years. It's already true
> in some parts of mathematics; a lot of papers published
> today done by humans are already obsolete and can be
> done using algorithms. Some of the problems we do today
> are completely uninteresting but are done because it's
> something that humans can do.

That's a pretty depressing assessment of the state of the field. But is it really true? I certainly have felt there are papers going into the journals that are just there because of the need to generate publications. But that's not always a bad thing. The unexpected consequences of doing something just for the sake of doing it have proven that untargeted driven research is sometimes the best way to glean genuinely new insights.

Like many colleagues, Jordan Ellenberg sees a vital role for humans in the future of our field:

> We are very good at figuring out things that computers
> can't do. If we were to imagine a future in which all the
> theorems we currently know about could be proven
> on a computer, we would just figure out other things
> that a computer can't solve, and that would become
> 'mathematics'.

But a lot of this human output is moving sideways rather than forward. We really are reaching the point in some areas where to go beyond the heights of Everest is going to necessitate getting in a machine. That's a shock for the old guard (and I probably include myself in there). That pen and paper will no longer hack it as a way to do ground-breaking mathematics is something that they are very reluctant to admit.

Voevodsky's visions

One of those who made his name in mathematics with pen and paper but has gone on to champion the importance of adding the computer to the armoury of the mathematician is Vladimir Voevodsky, one of the stars of my generation. I met him in Oxford when we were trying to tempt him with a position. People had spotted that he was a dead cert for a Fields Medal, and Oxford decided to get in early with a tempting offer. The seminars he gave on his work suggested a truly new vision of mathematics. This was not incremental advance or an interesting new fusion of established ideas. Voevodsky seemed to channel a new language of mathematics and was able to prove things that had eluded generations of mathematicians.

I spoke earlier in the book about three sorts of creativity: exploratory creativity, combinational creativity and transformative creativity, changing the landscape of a field by introducing a completely new perspective. Voevodsky's creativity was truly transformative. Listening to his ideas you couldn't help thinking: 'Where did that come from?'

It turned out that this exceptional creativity was enhanced by a rather unexpected source. I was quite shocked to learn during his visit that one of the important considerations as he considered his future place of work was access to drugs. And I'm not talking about caffeine, most mathematicians' drug of choice. (As the famous Hungarian mathematician Paul Erdös quipped: 'A mathematician is a machine for turning coffee into theorems.') We were asked to source some pretty hardcore class B drugs to convince him of Oxford's credentials.

I've never really felt that drugs would be much good in helping me access ideas which require a cold steely logic to be navigated, but Voevodsky felt that amphetamines could lead him to churn out visions that he could then check once he'd touched base again. I began to think he might well have been on to something when I saw what effect caffeine and amphetamines have on spiders building webs. Spiders on speed create fast coherent webs while the webs of caffeinated spiders are a total mess. Voevodsky went on to win his Fields Medal and accepted a position at the Institute for Advanced Study in Princeton, but his early successes triggered something of an existential crisis.

'I realised that the time is coming when the proof of yet another conjecture won't have much of an effect,' he said. 'I realised that mathematics is on the verge of a crisis, or rather, two crises.'

The first of these two crises involved the separation of 'pure' and 'applied' mathematics. As budgets for research are increasingly squeezed, governments are having to make hard choices about where money should be spent. Some politicians are beginning to question why society should pay money to people who are engaged in things that do not have any practical applications. Voevodsky felt it was important to show why the very esoteric research that he was doing could nonetheless have enormous practical impact on society.

But it was the second crisis that was more of an existential threat, and it related to the increasing complexity of pure mathematics. Even if mathematicians were able to master their little corner, it was becoming impossible for the community to verify others' work. Mathematicians were becoming more and more isolated. Already in 1739 David Hume pointed out in his *Treatise on Human Nature* the importance of the social context of proof:

> There is no Algebraist nor Mathematician so expert in
> his science, as to place entire confidence in any truth
> immediately upon his discovery of it, or regard it as
> anything, but a mere probability. Every time he runs over
> his proofs, his confidence increases; but still more by the
> approbation of his friends; and is rais'd to its utmost

perfection by the universal assent and applauses of the learned world.

Sooner or later, Voevodsky believed, the journal articles would become too complicated for detailed verification to take place and this would lead to undetected errors in the literature. And since mathematics is a deep science, in the sense that the results of one article usually depend on the results of many previous articles, this accumulation of errors would be very dangerous.

Having identified these two potential crises, Voevodsky decided to leave the research that had won him fame and glory and to focus his work on averting the potential catastrophes facing the field of mathematics. He began with his first challenge of applying his mathematics to solve problems in other fields. He had always been interested in biology since he was a kid, so he wondered whether the tools he had developed might provide new insights into a field that is generally regarded as very unmathematical. He spent several years trying to determine whether you could deduce the history of a population by analysing its current genetic make-up. But his attempts to crack this biological riddle eventually ran aground. He found he didn't have the same tools and skills to dig deep into biological questions as he did in his chosen area of mathematics.

'By 2009, I realised that what I was inventing was useless. In my life, so far, it was, perhaps, the greatest scientific failure. A lot of work was invested in the project, which completely failed.'

After much soul searching he turned to the second crisis he'd identified: the increasing complexity of cutting-edge mathematics. If humans were unable to check each other's proofs then maybe we needed to enlist the help of machines. For a pure mathematician of Voevodsky's calibre to start talking about using computers seemed to many a misguided move. Most mathematicians continued to believe in the power of the human mind to navigate equations and geometries and, guided by their aesthetic sensitivity, to sniff out solutions. But those who were critical of his decision did not believe or appreciate that a crisis was imminent.

As Voevodsky looked around for suitable tools, he could see that the only viable computer project able to navigate proofs was the French system Coq. Initially he just couldn't get his head around how it worked. So he went back to basics and proposed to the Institute for Advanced Study that he teach a course on Coq. This is a trick I've often used: if you don't understand something, try teaching it. Gradually it began to dawn on him that the language computer scientists were using, which initially appeared so alienating, was in fact just a version of the very abstract world in which he had spent his early years as a mathematician.

It was as if he'd managed to solve two crises at once. First, his obtuse mathematical ideas were perfect for articulating the very practical world of modern-day computing and, second, here was a new language with which he could build a new foundation for mathematics where the computer would play a central role.

Voevodsky's vision of the future of mathematics is far too revolutionary for most mathematicians, many of whom believe he has moved to the dark side. There is still a deep divide between those doing mathematics with pen and paper (perhaps using a computer now and again to check a routine calculation) and those who want to use computers to prove new theorems. The idea of using computers to check proofs is becoming acceptable: the human who created the proof is still in the driving seat here. It's when it comes to computers actually creating the mathematics that people, myself included, start to have issues.

But Voevodsky believes these old attitudes will have to be dropped: 'I can't see how else it will go. I think the process will be first accepted by some small subset, then it will grow, and eventually it will become a really standard thing. The next step is when it will start to be taught at math grad schools, and then the next step is when it will be taught at the undergraduate level. That may take tens of years, I don't know, but I don't see what else could happen.'

Voevodsky has compared the interaction to playing a computer game. 'You tell the computer, try this, and it tries it, and it gives you back the result of its actions. Sometimes it's unexpected what comes out of it. It's fun.'

Voevodsky never had a chance to see how his revolution would pan out. Tragically he died in 2017 from an aneurysm at the age of fifty-one.

So it is in the spirit of Voevodsky's vision that I have decided to embrace the future and be open to the potential of the computer to extend mathematical creativity. Given the close link that has always existed between mathematics and music, I wondered if I might get some inkling of the role that the computer might play in doing mathematics by looking at how AI has extended musical composition. After all, as Bach's student Lorenz Mizler von Kolof once said: 'Music is just the process of sounding mathematics.'

11

MUSIC: THE PROCESS OF SOUNDING MATHEMATICS

*Music is the pleasure the human mind experiences from
counting without being aware that it is counting.*
Gottfried Wilhelm Leibniz

When Philip Glass went to study with Nadia Boulanger in Paris
in 1964, every lesson began with Bach. *The Art of Fugue* was a
key part of the curriculum, and Glass was made to learn a new
Bach chorale each week. Once he'd mastered the chorale, a hymn
based on four voices, he was instructed to add four new voices on
to the original four in such a way that no voice repeated another
and yet they all meshed seamlessly. Boulanger believed all great
composers had to start by immersing themselves in Bach.

I think a small part of me wishes I had been a composer rather
than a mathematician. Music has been a constant companion on
my mathematical journeys. My mind searches for patterns and
structures as I contemplate the unexplored reaches of the mathematical landscape, which may be why a soundtrack by Bach
or Bartók helps my thought process. Both were drawn to structures that are similar to those I find exciting as a mathematician.

Bach loved symmetry. Bartók was fascinated by the Fibonacci numbers. Sometimes composers are intuitively drawn to mathematical structures without realising their significance; at other times they seek out new mathematical ideas as a framework for their compositions.

It was while talking to Emily Howard, a composer, about geometric structures that might be interesting to explore musically that I had an idea. Perhaps, in exchange for a tour of the mathematics of hyperbolic geometry, she might agree to give me composition lessons. She thought this was a fair deal and we met over coffee not long after that for my first lesson.

Just as a blank piece of paper can present a daunting void for a novice writer, the sight of a musical stave with no notes put me into a panic. Emily explained calmly that every composer needs to start with a framework or set of rules to help give shape to their composition. She suggested we start with the rules governing medieval polyphony, where something called a prolation canon is used to take one line of music and grow it into a multi-voiced work. The idea is to start with a simple rhythm that will be sung by one voice. Then a second voice sings the same rhythm at half-speed and a third at twice the speed. In this way you get three voices singing different rhythms that are nonetheless strongly correlated. When you listen to a piece of polyphony using this technique, your brain recognises that there is a pattern connecting the three voices.

Here was my homework: to compose a simple rhythm and grow it into a string trio using the medieval tradition of prolation. This was a simple enough undertaking, and one that can easily be mapped out as a mathematical equation: $x + 2x + \frac{1}{2}x$. As the piece I composed emerged, I had a strong sense of being a bit like a gardener. I began with a small fragment of rhythm that I had created from nothing. This was like a seed that I then threw down on to the stave. But then, by applying the algorithm that Emily had given me, I could take this seed, mutate it, change it, grow it, and the algorithm would start to help me to fill out the rest of the stave with bits of music that had a strong connection with the original seed but weren't simply repetitions of the same bit of music. It was a deeply satisfying experience to see my musical garden growing out of this simple rule.

It was composing this simple piece that helped me to understand the close correlation between algorithms and composition. An algorithm is a set of rules that can take different inputs and by applying the rules on the input lead you to a result. The initial input is the seed. The algorithm is the way to grow that seed. We've seen algorithms that take two numbers, and by applying Euclid's algorithm you can discover the largest number that divides both the original numbers. There are algorithms that take different images and, by analysing the picture, can tell you what is in the image. There are algorithms that grow fractal graphics: by starting with a simple geometric image and by repeatedly applying a mathematical equation to the image a complex graphic emerges.

The algorithms that apply to music have a similar quality. One of Philip Glass's early pieces perhaps illustrates why algorithms are a key device in the composer's toolbox. Called *1 + 1*, the piece is for a single player, who taps out a rhythmic sequence on a tabletop amplified via a contact microphone. The seeds for the piece are two rhythms: the first rhythm, which I'll call A, is made up of two short beats followed by a long beat; the second, called B, is just a single long beat. Glass then instructs the player to combine the two units using a choice of some regular arithmetic progressions. This is the algorithm that grows the seed.

The performer is given the freedom to choose their own algorithms, but Glass gives some examples of different arithmetic progressions that can be used to grow the piece. For example, ABAABBBAAABBBBB . . . So the A rhythm increases by one each time but the B increases by two. I think a lot of people have criticised Glass, saying: 'Come on, where is the music here, it just sounds monotonous', but to me this piece crystallises what's at the heart of all music – as you listen, your brain recognises that the piece isn't random and nor is it simple repetition. There's a pleasure in trying to reverse-engineer the construction of the piece and spot the patterns underpinning it. And it's this idea of pattern that I believe connects music so closely with the world of mathematics.

Part of the art (or perhaps science) of the composer is therefore twofold: coming up with new algorithms that might be used to create interesting music and then choosing different seeds of music that can be fed into the algorithm. Given that there is this algorithmic quality at work that is growing the music, could this be the key to how a computer might embark on becoming a composer?

Bach: the first musical coder

One of the reasons why Boulanger had insisted Philip Glass take Bach as his starting point for musical composition is that algorithms are very much in evidence in the way that Bach creates his music. In some ways I think that Bach deserves the title of being one of the first musical coders (that's coders, not codas!). They are more complex than the simple algorithm behind medieval polyphony, but many of his compositions could be mapped out in mathematical terms. The *Musical Offering*, inspired by a challenge set by Frederick the Great, illustrates this most clearly.

Although the Prussian king is best known for his military victories, Frederick the Great had nevertheless been passionate about music all his life. Despite his father's attempts to literally beat such frivolous pursuits out of him as a child, Frederick was happy to combine his military prowess with a celebration of the best musical talents in his court in Potsdam. Among them was Bach's son Carl Philipp Emanuel, employed as chief harpsichordist.

The *Musical Offering* grew out of a visit that Bach senior

made in 1747, aged sixty-two, to his son while he was serving at the court. The trip had taken the old Bach several days of hard travelling, and when he arrived he was looking forward to collapsing at his son's house. However, when Frederick the Great was brought the list of strangers that had arrived in town that night, he declared excitedly, 'Gentlemen! Old Bach is here!' and immediately sent out a request for Bach to join him for an evening of music making. He was particularly keen to show off his new collection of fortepianos. It is said that he had been so impressed by the pianos of Silbermann of Freyberg that he had bought all fifteen pianos they had made and they were scattered around the palace.

Having received the summons from the palace, Bach did not even have time to change from his travelling clothes. You did not keep the king waiting. On his arrival, they moved from room to room trying out the pianos. Having heard about Bach's fantastic abilities to improvise on the spot, Frederick the Great sat down and challenged Bach to create a piece based on a theme that the king tapped out on his new fortepiano.

This was no ordinary tune. It was full of chromatic steps without any clear key. It was impossibly long and complex. Indeed, the twentieth-century composer Arnold Schoenberg marvelled at how cleverly it had been constructed so that it 'did not admit one single canonic imitation'. In other words, it was resistant to any of the Classical rules of counterpoint. Indeed, some have suggested that Frederick the Great had cooked up this impossible challenge with Bach's son. C. P. E. Bach was quite tired of living in the limelight of his father. He regarded his father's work as old school and wanted to write a new style of music. So perhaps the challenge was meant to reveal the shortcomings of his father's style and method. As Schoenberg said, they wished 'to enjoy the helplessness of the victim of his well-prepared trap'. If that's so, it backfired spectacularly. The old Bach sat down and proceeded to improvise the most stunning three-part fugue based on this tricky theme.

A fugue is a more sophisticated version of a canon or round, something many people sing in school. In a canon one half of the class starts singing a song and then a little later the second half start singing the same song. The art of a good canon is creating a song that, when shifted in time, sits nicely on top of the original tune to harmonise it. 'London's Burning' or 'Frère Jacques' are the most obvious examples.

The algorithm at work here is quite simple and has a very geometric quality. First of all create the tune that will be the basis for the canon. Write the tune out on the musical stave. Now the algorithm is a rule that gets applied to this input to produce a piece full of harmony. The way the algorithm works is that it takes a copy of the original tune and then repeats the same tune but shifted a certain number of notes to the right. This has the effect of shifting it in time. It's a bit like a frieze pattern on a pot that gets copied, shifted and repeated. Just as with a pot, you can shift the tune again, creating a third voice which sings the tune after the first two voices have started.

If one wanted to try to write the canon algorithm as a mathematical formula, then it takes a tune, X, and then a choice of time delay, S, and then plays X + SX + SSX. The algorithm creates a harmonised piece with three voices out of a single tune.

A fugue develops this further, with multiple voices and variations on the themes evolving throughout the piece. Another rule that Bach enjoyed applying to the original tune was to make the second voice not only shift to the right but also to shift up and down, changing the pitch. He also applied rules of symmetry to the tune. The second voice might play the tune backwards. This is like reflecting the pattern in a mirror. It was by combining all these rules that one could build an algorithm that started with a tune such as the one Frederick the Great had challenged Bach with and then produced a piece full of harmony and complexity. Bach had understood how this algorithmic approach to the king's challenge could help him to solve the puzzle of improvising a fugue on the spot.

Frederick the Great was impressed by the improvisation but would not stop there. Now he wondered if Bach could double the voices and improvise a fugue for six voices. This had never been done before. Still, the composer was not going to turn down the challenge without a fight. Six voices would require a bit more thought than simply sitting at the keyboard and improvising, so he went away to see if he could weave together six voices into a coherent fugue. The result was a stunning piece, now called the 'Ricercar', which he delivered to the king two months later.

Together with this fugue Bach composed ten pieces based on the theme Frederick had tapped out. In each piece he provided a simple tune and a mathematical rule or algorithm for expanding this tune into a harmonised piece. Each offering was presented as a puzzle that the performer would have to crack in order to perform it. For example, in one piece he writes a single line of music with an upside-down clef at one end. This upside-down clef is the key to the algorithm that Bach wants the player to apply to the tune. The algorithm says that you need to take the original piece of music and turn it upside down, and this upside-down piece is then played simultaneously with the original tune to create a piano piece for two hands. The algorithm is the rule that is applied to the original tune to make additional voices in the music. Just as an algorithm for identifying an image in a photo applies whatever the photo, this musical algorithm would create a piece whatever tune you gave it.

Each of the ten pieces that make up the beginning of the *Musical Offering* has its own different algorithmic tricks at work that mathematically transform the original theme. The ten pieces act as a warm-up for the extraordinary final fugue, which is a perfect illustration of how Bach can take a simple theme and, by applying simple mathematical algorithms, create a piece of exquisite complexity. The tune shifts in time, gets played backwards, climbs higher in pitch, gets turned upside down. It this dizzying mixture of rules that are combined so skilfully by Bach to produce the six-part fugue. Our brain responds to that tension between recognising a pattern and knowing it is not so simple that we can predict what will happen next. It is that tension between the known and the unknown that excites us. As Harrison Birtwistle put it: music shouldn't finish before it has ended.

Was Bach aware of all the mathematical games he was playing? For me it is clear he knew what he was doing. There are too many examples of mathematical structures that it would be hard to put in accidentally or even subconsciously. He was a member of the Corresponding Society of Musical Sciences founded by his student Mizler. The society was dedicated to exploring the connections between science and music and circulated papers with titles like 'The Necessity of Mathematics for the Basic Learning of Musical Composition'. So Bach was certainly immersed in a world that was interested in the dialogue between mathematics and music.

Bach's son Carl Philipp Emanuel was rather dismissive of his father's fugues, declaring himself 'no lover of dry mathematical stuff'. To try to prove that there wasn't really much more than musical trickery at work, he even prepared a musical parlour game which he entitled 'Inventions by Which Six Measures of Double Counterpoint Can be Written without Knowledge of the Rules'. Players were handed two sheets of music. Each page consisted of what looked like a random selection of notes. The first page would be used to write music for the right hand, the descant. The second page would create the bass line for the left hand. All the player had to do was to pick a note at random from which to start the tune and then play the ninth note after that then the eighteenth note, the twenty-seventh note, until the notes ran out. C. P. E. Bach's skill was in choosing notes so that wherever you started by following this simple rule of playing every ninth note, anyone could build up a piece of acceptable counterpoint without having a clue as to how it was done. Perfect code for a machine!

While the *Musical Offering* is often performed in concerts, I have never heard of his son's 'Inventions' being played, suggesting that there may be more to successful musical composition than mechanically following a set of rules.

Mozart too is credited with a similar algorithm to C. P. E. Bach's game to allow players to manufacture their very own Mozart waltz. His *Musikalisches Würfelspiel*, or 'musical dice game', is a way of generating a sixteen-bar waltz using a set of dice. It was first published in 1792, a year after Mozart's death. Some have questioned whether the game was actually cooked up by the publisher Nikolaus Simrock, who just put Mozart's name to the game to boost sales.

The game consists of 176 bars which are arranged in a 11x16 configuration. The first column gives you a choice of eleven different bars that can start the music. The way you choose which bar to start with is by throwing two dice and subtracting one from the resulting score, giving you a number between 1 and 11. So, for example, if I throw a double six, then that means I start my piece by playing the eleventh bar in the first column. The second column controls the second bar and again the choice of which of the eleven bars you play is determined by throwing the two dice again. In this way you run through the sixteen columns, throwing the two dice each time to pick out which of the eleven bars will be played.

The staggering thing is that there are 11^{16} different waltzes that you can generate using this system. That's a total of 46 million billion waltzes. Played one after the other, it would take 200 million years to hear every waltz. The element of randomness combined with some predetermined structural elements is a trick that early algorithmic artists would later use. The mastery of Mozart's composition is to produce 176 bars that fit together into a pretty convincing waltz whatever the throw of the dice. Inevitably not all variations are pleasing to the ear. Some combinations work better than others. For me, this is one of the problems with such open-ended algorithms. There is a frustration at the fact that Mozart hasn't curated which waltzes work better than others.

Emmy: the AI composer

I enjoy challenging myself to name the composer of the pieces I hear on the radio before finding out who it is. As I listened one morning, working away at my desk, I quickly homed in on Bach as the most likely candidate for the piece that was playing. When it came to an end I was in for a shock: the presenter revealed that the piece had been created by an algorithm. I think what shook me was not that I had been duped into thinking it was Bach so much as the fact that during the short period I'd been listening to the musical composition, I'd been moved by what I'd heard. Could a bit of code really have done that? I was intrigued to find out how the algorithm behind it had tricked me into thinking the great Bach had composed it.

Bach is the composer most composers begin with, but he is the composer most computers begin with too. The piece I listened to on the radio that day was generated following simple rules of code cooked up by a composer who had been struggling for inspiration. David Cope first turned to algorithms out of desperation. He'd been commissioned to write a new opera and was procrastinating, unable to commit notes to stave. But then he had an idea. He remembered how Ada Lovelace had speculated that 'the engine might compose elaborate and scientific pieces of music of any degree of complexity or extent' and decided to explore her idea.

He started experimenting by feeding punch cards into an IBM computer (this was the early 1980s). Notes appeared as output. His early experiments, he later admitted, were truly dreadful. But he persevered, heading off to Stanford to do a course in computer music. With the deadline for his opera commission fast approaching, he decided to put his computing skills to the test.

If he could get an algorithm to understand his compositional style, then whenever he got stuck and didn't know how to proceed, the algorithm could make a suggestion that would be compatible with his particular way of composing. Even if the algorithm suggested something Cope considered absurd, it would at least help him understand what might have been a better choice. The algorithm would act as catalyst to spur his creativity. Cope called his new concept 'Experiments in Musical Intelligence', or EMI for short. The alter ego composer that began to emerge from these algorithmic experiments would later be called Emmy, partly to avoid confusing the project with the British recording label EMI, partly to give the algorithm a more human name.

Having wrestled for seven years trying to write his opera, with the help of Emmy, Cope completed the piece in two weeks. Calling it *Cradle Falling*, he decided at that stage not to let on that a computer had helped him compose the piece so as not to bias the critics' reaction. When it was premiered two years later, in 1987, Cope was amused to find that the piece garnered some of the best reviews of his career to date. One critic declared: 'It was most moving. "Cradle Falling" unquestionably is a modern masterpiece.' The reaction inspired Cope to continue his collaboration with Emmy.

If the algorithm could learn Cope's own compositional style, could it be trained on more traditional composers? Could it for instance take the compositions of Bach or Bartók and write pieces that they might have written? Cope believed that every piece of music had encoded within it instructions to create other pieces that were similar but subtly different. The challenge was to figure out how to crystallise these instructions into code.

He began, with Emmy's help, to build up for each composer a database of ingredients that would correspond to their particular style, like the vocabulary and grammar of their own musical language. The notes were the letters, but what words would correspond to the language specific to this particular composer? One of the key concepts behind Cope's analysis was the idea of the signature motif, a sequence of four to twelve notes that can be found in more than one work by the same composer. In Mozart's piano concertos, for example, one can find a phrase used over and over, which is called an Alberti bass pattern. It is often in the second line of music and consists of three notes played in a sequence of 13231323.

This pattern would go into the database corresponding to Mozart's style. Of all the composers Cope analysed, Mozart's was particularly signature-rich. The signatures might occur at different speeds and pitches, but mathematics is very good at spotting the underlying pattern. It's a bit like recognising that even though you might throw a ball through the air in many different ways, the ball will always follow a path described by a parabolic equation.

Cope's analysis revealed strong patterns across a composer's output. From Bach to Mozart, Chopin to Brahms, Gershwin to Scott Joplin, each has a particular pattern of notes that they seem to be drawn to. Perhaps this shouldn't be surprising. Why, after listening to a couple of bars on the radio, will I so often be able to recognise a composer even if I've never heard the piece before? Like a blind wine taster I'm tapping into key indicators, which in the case of music will be patterns of notes. They are like the painter's signature brushstrokes. Some composers, like Bach, went so far as to sign their names in notes. The final fugue of the *Art of Fugue* features the notes B flat, A, C and B, which in German notation is represented by the letters B A C H.

Having cut up the pieces into cells and signatures to form a database for each composer, Cope's algorithm now turned to what he called 'recombinance'. It is one thing to recognise the building blocks of a complex structure and another to find a way to construct a new composition from the pieces. Cope could have chosen to use a random process like Mozart's game of dice. But randomly combining fragments is unlikely to mirror the emotional tension and release that a composer builds into a composition. So he added one more step to his program: he created a heat map for every piece.

Composers will often put together elements to build up a grammar which musicians call a phrase. These frequently have a pattern that Cope tried to abstract in a shorthand he called SPEAC. If the database is the dictionary, then SPEAC corresponds to the way in which the composer uses the words in the dictionary to write sentences. SPEAC identifies five basic building blocks for musical phrases:

(S)tatement: musical phrases that 'simply exist "as is" with nothing expected beyond iteration'.

(P)reparation: these 'modify the meanings of statements or other identifiers by standing ahead of them without being independent'.

(E)xtension: a way to prolong a statement.

(A)ntecedent: phrases that 'cause significant implication and require resolution'.

(C)onsequent: These resolve the antecedent. 'Consequents are often the same chords or melodic fragments as found in S. However, they have different implications.'

Many Classical composers use this grammar, sometimes unknowingly, but often they will have been taught it as part of their training. Certain chords feel like they need to go somewhere to be resolved. The chords that follow can make you feel you have arrived home or can further crank up the need for resolution. SPEAC helped Cope analyse the ebb and flow of a piece. Each composer had their own particular version of this grammar. Here, for example, is Cope's analysis of one of Scriabin's piano pieces:

Once he had established this basic grammar, Cope measured the tension created by the use of certain intervals. If you have an octave interval or a perfect fifth, this does not cause great tension, a fact that is reflected by the mathematics. These are intervals where the frequencies are in small whole number ratios: the octave is 1:2; the perfect fifth is 2:3. However, if you have an interval where two notes next to each other on the piano are played together (a semi-tone or minor second), then the notes sound like they are clashing. There is a high degree of tension. Again, the mathematics reflects this: the frequencies are in a ratio corresponding to much bigger numbers (15:16). If you hear these high-tension intervals in a piece of music, they will generally be followed by a move to head towards low-tension resolutions.

These rules were fed into the system to help build a new composition from the large database of a given composer's signatures. Emmy's recombination rules take fragments and hook them together following certain guiding principles. For example, fragment A can be followed by fragment B if fragment B begins in the same way as fragment A ends but now heads off in a new direction. The fragments have to match the grammar encoded by Cope's SPEAC analysis.

When many different fragments would fit, a choice needs to be made. Cope is not a fan of using randomness as a way to generate choices. He prefers to use a choice of mathematical formula which provides an arbitrary structure to control the choices made, much like the 'unaccountable predictability' guiding the Painting Fool. By 1993 Cope and Emmy were ready to release their first album of pieces created in the style of Bach called *Bach by Design*. The pieces were quite tricky and he failed to find a human to perform them, so he had to resort to a computer being both composer and performer. The album was not well received by critics.

'When I read the reviews, I was very upset that they weren't primarily about how the pieces were composed, they were primarily about how they were performed.' Given that the composition hadn't been attacked, he felt emboldened to continue the project, producing a second album in 1997 with pieces in the style of some of the other composers he'd analysed: Beethoven, Chopin, Joplin, Mozart, Rachmaninov and Stravinsky. This time the pieces were performed by human musicians. The critics' response was much more positive.

'The Game': a musical Turing Test

But would the output of Cope's algorithm produce results that would pass a musical Turing Test? Could they be passed off as works by the composers themselves? To find out, Cope decided to stage a concert at the University of Oregon in collaboration with Douglas Hofstadter, a mathematician who wrote the classic book *Gödel, Escher, Bach*. Three pieces would be played. One of these would be an unfamiliar piece by Bach, the second would be composed by Emmy in the style of Bach and the third would be composed by a human, Steve Larson, who taught music theory at the university, again in the style of Bach. The three pieces would be played in random order by Larson's wife, Winifred Kerner, a professional pianist.

Larson was upset when the audience declared his two-part invention à la Bach to be the one composed by a heartless computer. His disappointment was soon eclipsed, however, by the shocking vote for the algorithmic Bach over the great man himself. The real Bach was judged a poor imitation!

'I find myself baffled and troubled by Emmy,' Hofstadter mused, in trying to make sense of the results. 'The only comfort I could take at this point comes from realising that Emmy doesn't generate style on its own. It depends on mimicking prior composers. But that is still not all that much comfort. To what extent is music composed of 'riffs', as jazz people say? If that's mostly the case, then it would mean that, to my absolute devastation, music is much less than I ever thought it was.'

Cope went on to repeat 'The Game' in a number of other venues around the world. The audience's reactions began to unnerve him. In Germany a musicologist was so incensed he threatened Cope following the concert, declaring that he had killed music. The musicologist was quite large and about 45 kilos heavier than him, and Cope felt only the crowd surrounding him had protected him from a punch-up. At another concert Cope recalled how a professor came up at the end of the performance and told him how moved he had been. 'The professor came to me and said this was one of the most beautiful pieces he'd heard in a long time.' He didn't realise until the lecture following the

concert that the music had been composed by a computer algorithm. This new information totally transformed the professor's
impression of the work. He found Cope again after the lecture
and insisted on how shallow it was. 'From the minute it started
I could tell it was computer-composed,' he now said. 'It has no
guts, no emotion, no soul.' Cope was stunned by the totality of
his reversal. The output was the same: the only thing that had
changed was his knowledge of the fact that it had been generated
by computer code.

On another occasion, when Hofstadter played two pieces,
one by Chopin and the other a Chopin-like piece composed by
Emmy, an audience made up of many composers and musical
theorists were duped into believing that the computer-generated
piece was the real one. A member of the audience wrote to Cope
in admiration afterwards, describing her shock upon learning she
had voted the wrong way: 'There was a collective gasp and . . .
what I can only describe as delighted horror. I've never seen so
many theorists and composers shocked out of their smug com-
placency in one fell swoop (myself included)! It was truly a thing
of beauty.'

Hofstadter was genuinely surprised by the Chopin-esque piece
produced by Emmy: 'It was new, it was unmistakably Chopin-like
in spirit, and it was not emotionally empty. I was truly shaken.
How could emotional music be coming out of a program that
had never heard a note, never lived a moment of life, never had
any emotions whatsoever?'

Cope believes his algorithm is so successful because it gets to
the heart of how people write music. 'I don't know of a single
piece of expressive music that wasn't composed, one way or
another, by an algorithm,' he says. While this may strike listeners
as a baffling or even offensive statement, many composers would
agree. It is only those on the outside who daren't admit that their
emotional state can be pushed and pulled around by code. 'The
notion that humans have some kind of mystical connection with

their soul or God, and so on, allowing them to produce wholly original ideas (not the result of recombination or formalisms), seems ridiculous to me,' Cope confided.

This may well be, but I think it is important to recognise that although music may be more mathematical and coded than we generally acknowledge, that does not rob it of its emotional content. When I speak about the connections between mathematics and music, people will sometimes get quite upset, imagining that I am making the music they love into something cold and clinical. But this misses my point. It isn't so much that music is like mathematics as that mathematics is like music. The maths we enjoy and are drawn to has huge emotional content. Those who can appreciate the language of maths will be pushed and pulled by the twists and turns of a proof just as so many of us are moved when we listen to a piece of music unfolding.

I think the human code running in our brains has evolved to be hyper-sensitive to abstract structures underpinning the mess of the natural world. When we listen to music or explore creative mathematics, we are being exposed to the purest forms of structure and our bodies respond emotionally, to mark the recognition of this structure against the white noise of everyday life.

What accounts for the difference we perceive between a random sequence of notes and a sequence we regard as music? According to the work of Claude Shannon, the father of information theory, part of our response comes down to the fact that a non-random sequence has some algorithm at its base that can compress the data, while the random sequence does not. Music is distinct from noise by virtue of its underlying algorithms. The question is which algorithms will make music that humans feel is worth listening to?

Many people will not give up on the idea that music is at some level an emotional response to life's experiences. These algorithms are all composing in soundproof rooms with no interaction with the world around them. Without embodied experience one cannot hope to emulate the music of the greats. Hofstadter certainly believes – or maybe hopes – that this is the case:

> a 'program' which could produce music as [Chopin or
> Bach] did would have to wander around the world on
> its own, fighting its way through the maze of life and
> feeling every moment of it. It would have to understand
> the joy and loneliness of a chilly night wind, the longing
> for a cherished hand, the inaccessibility of a distant town,
> the heartbreak and regeneration after a human death.
> Therein, and therein only, lie the sources of meaning
> in music.

But it is the listener who brings their emotions to the music. The role of the listener, viewer or reader in creating a work of art is often underestimated. Many composers argue that this emotional response emerges from the structure of the music. But you don't program in emotion. Philip Glass believes emotions are generated spontaneously as a result of the processes he employs in his compositions. 'I find that the music almost always has some emotional quality in it; it seems independent of my intentions.'

The relationship between music and emotions is one that has long been a source of fascination for composers. Stravinsky, whose compositions are so expressive, was particularly eloquent on the subject. He believed that the emotions belonged not to the music but to the listener:

> music is, by its very nature, powerless to *express*
> anything at all, whether a feeling, an attitude of mind,
> a psychological mood, a phenomenon of nature . . . if,
> as is nearly always the case, music appears to express
> something, this is only an illusion and not a reality. It
> is simply an additional attribute which, by tacit and
> inveterate agreement, we have lent it, thrust upon it, as

a label, a convention – in short, an aspect unconsciously or by force of habit, we have come to confuse with its essential being.

Why, then, does music seem to illicit such powerful emotional responses? Perhaps composers have managed to identify the way the brain encodes certain emotions. These frequencies or notes coding emotions may be different for different people. Most people will agree that a certain sequence of notes in what we call a minor scale is associated with sadness. Is that a learned or innate response? A composer may choose a minor key to capture a mood and that suggests a direct encoding, but music theory has not advanced to the stage where we understand a huge amount about how this encoding works. So a composer is probably working in the dark, much as Stravinsky and Glass suggested: they create structure and the emotion emerges from the structure.

Many composers like to set up rules or structures to help them generate musical ideas. Bach enjoyed the puzzle of writing fugues. Schoenberg initiated a whole new school of composition around themes that included all twelve notes of the chromatic scale. Bartók was driven to create works that grew in tandem with the Fibonacci numbers. Messiaen used prime numbers as a framework for his *Quartet for the End of Time*. And Philip Glass eventually emerged from his torturous apprenticeship with Nadia Boulanger to create the additive process, from which his distinctive minimalist music emerges.

Stravinsky believed that constraints were key to his creativity:

My freedom consists in my moving about within the narrow frame that I have assigned myself for each one of my undertakings. I shall go even further: my freedom will be so much the greater and more meaningful the more narrowly I limit my field of action and the more I surround myself with obstacles.

My composition tutor had set me off on my own small musical journey with a set of rules to assist me. After starting with prolation canons I went on to create some of my own constraints and came up with a few algorithms to guide me in my composition. I'd read that John Cage would often compose a piece on the page without really knowing what it would sound like until it was played for the first time, and I was curious to hear what my mathematical reimaginings would sound like.

But when I sat at the piano and sounded out the string trio I'd composed, I was disappointed. The rules I'd followed meant that the piece had an interesting logic to it that took the listener on a journey, and yet it didn't sound right. I don't really know what that means, and of course it's silly to suggest that music has a right or a wrong answer like mathematics, but having been disappointed by the initial results I began to break the rules I'd set up, to perturb the notes I'd penned on the page to create something that made more musical sense to my ears. I can't really articulate why I made the changes I did as I allowed myself to be guided by something deeper, the relationship of my physical body to the music, my subconscious, my humanity.

This was an important lesson. Composition is a fusion of rules and patterns and algorithms and something else. That something draws on all those things Hofstadter believes we get by wandering around the world. It is that mysterious something else which was starting to bleed into my notes that began to give it life and beauty.

Do these structures need to be informed by an awareness of emotions? If so, how could a computer ever hope to gain this awareness? If music is encoding emotions, could that code be used to simulate an emotional state in the computer? Perhaps the 20,000 lines of code that created Emmy is already part-way there. If Hofstadter is having an emotional reaction to the Chopin produced by Emmy, then isn't this really an emotional reaction to 20,000 lines of code? Hasn't that code captured emotion in just the same way as the notes on the stave that Chopin himself wrote?

To call Emmy's output music composed by AI is something of a con. Emmy is dependent on having a composer prepare the database. It relies on composers of the past having created sound worlds that it could plunder. Cope, as a composer, had the analytic tools and sensitivity to pick out the elements that correspond to a composer's style and the skill to figure out how those elements should be recombined. Much of Emmy's creativity comes from Cope and from the back catalogue of the musical greats of history.

Cope built Emmy using a top-down coding process: he wrote all of the code to output the music. We are now at a point where new and more adaptive algorithms could be exposed to the raw data of a composer's scores and trained to learn musical theory from scratch, without passing through the filter of human musical analysis. So will machine-learning algorithms be able to create Classical compositions that rival the greats from scratch? The answer, as is so often the case in musical theory, takes us back to Bach.

DeepBach: re-creating the composer from the bottom up

Bach wrote 389 chorales, hymn-like pieces for four voices that Glass was asked to enhance and Cope analysed by hand. His famous *St John Passion* includes several chorales that punctuate the oratorio. If you are looking for an example of Bach's mathematical obsessions, you will find it here in the choices he has made. Bach was obsessed with the number 14. Many European thinkers and philosophers during this period were interested in the kabbalah, which involved changing letters into numbers and exploring the numerical connections between words to infer deeper connections, and Bach was intrigued to discover that when the letters of his surname (BACH) were translated into numbers and added up he got $2 + 1 + 3 + 8 = 14$. This became his signature number, a bit like the number that footballers wear on their shirts. Bach insisted, for example, on waiting to become the fourteenth member of the Corresponding Society of the Musical Sciences set up by his student Mizler. He also found interesting ways to introduce the number into his compositions. In the *St John Passion* we find eleven chorales. If you look at the number of bars in each of the first ten chorales they are as follows:

11, 12, 12, 16, 17, 11, 12, 16, 16, 17

The eleventh chorale that follows is the key here: it has 28 (or 2×14) bars. Now take the preceding chorales in pairs, starting with the first and tenth chorale: $11 + 17 = 28$. The second and ninth chorales are $12 + 16 = 28$. If you take the chorales in pairs in this symmetrical manner the bars always add up to 28. A coincidence? Hardly.

To compose these chorales Bach often starts with a Lutheran melody that forms the soprano part and then fills in the other parts to harmonise the melody. Cope programmed this harmonising into his algorithm by hand, based on his analysis of the chorales. He discerned the rules Bach was using to navigate his way through the harmony. But could a computer take the raw data and learn the rules of harmony for itself?

The exercise of harmonising a chorale is like playing a complex game of patience or doing an open-ended Sudoku puzzle. At each step you have to decide where your tenor voice is going to move next. Up? Down? How far up or down? How fast? This has to be done while taking into account the ways in which you are moving the other two voices you are weaving, and the whole thing has to underpin the melody.

When you are learning to do this exercise as a composition

student, there are a number of rules a teacher will impose. For example, you need to avoid two consecutive perfect fifths or octaves. A piece with consecutive fifths is felt to weaken the independence of the two voices and to degrade the harmonic effect. It's as if one channel has dropped out. The banning of parallel fifths was introduced as early as 1300 and remains a staple of compositional theory.

Glass recalls how, during one session, his teacher, Nadia Boulanger, started questioning his health: 'Not sick, no headache? Would you like to see a physician or a psychiatrist? It can be arranged very confidentially.' When he insisted he was fine she wheeled round in her chair, pointed at his chorale exercise for the week, and screamed: 'Then how do you explain this?!' Sure enough, Glass could see hidden fifths lurking between the alto and bass parts he had written.

It is the mark of the creative thinker to break with traditional rules. In AlphaGo we saw this in move 37 of game 2. Likewise we find Bach getting to the end of a chorale by sometimes breaking the rule of no parallel fifths. But does that make it a bad chorale? As my own tutor Emily explained to me, part of the joy of composing is to break these rules. That's your best chance of achieving Boden's idea of transformational creativity.

Harmonising a chorale has a two-dimensional quality to it. The harmony has to make sense in a vertical direction, yet the voices sung on their own, the horizontal direction, also have to have a logic and consonance to them. It is a testing challenge for human composers to write these chorales and get the two dimensions to coalesce.

So is this something a new algorithm driven by machine learning could engage with? Is the secret to Bach's skills decodable from his 389 published chorales? One approach to testing this proposition would be to do a statistical analysis to guess at the most likely direction a voice will head in, given what it has just sung. For example, you might see the sequence of notes ABCBA occur several times in various different chorales as part of one of the harmonising voices. You could then do a statistical analysis of the notes that follow the note A. In BWV 396 the next note descends to a G#. Yet if you take the data from BWV 228 the next note jumps up to an F. (BWV stands for *Bach-Werke-Verzeichnis*, which catalogues all of Bach's compositions.) By building up a statistical analysis you could create a game of musical dice with different weightings for different possible notes to continue the phrase. Let's imagine you get eight cases where Bach chooses the G# and four cases where he chooses F. So two out of three times you'll get the algorithm to go for the G#. It's a bit like the way DeepMind's algorithm learned to play Breakout: which way should the algorithm move the paddle, and by how much, to win the game? The paddle is replaced by a voice singing higher or lower notes.

The challenge with such an approach, as Cope discovered when he set out to identify a composer's signature phrases, is to determine how many notes to condition the next choice. Too few and you can go anywhere. Too many and the sequence will be over-determined and you start to copy what Bach's done. Then, alongside pitch, you have to factor in rhythm patterns.

Moving from left to right and building the voices out of what has gone before seems to be the most obvious approach, given that this is how we hear music. But it isn't the only way to statistically analyse a piece. DeepBach, an algorithm developed by a music student, Gaëtan Hadjeres, for his PhD thesis under the supervision of François Pachet and Frank Nielsen, seeks to analyse Bach's chorales by taking them outside time and viewing the chorales as two-dimensional geometric structures. If you remove a piece of the geometrical structure and analyse the surrounding image, you can guess at how Bach might have filled in the rest of the shape. So rather than composing forward in time, it looks at the parts threaded backwards. This is a typical trick in solving a puzzle: start at the end and try to work out how to get there. But one could also take middle sections and ask how Bach filled these in.

This multidimensional analysis led to a more structurally coherent chorale than those created by algorithms that meander forward without really knowing where they are heading, nudged on by what has happened before. Yet the analysis is still really being done on a local level. The algorithm looks at a ball around each note and tries to fill in the note based on the ball, but the size of the ball is constrained. In DeepBach's case it considers four beats on either side of a given note. So how successful is the algorithm?

Gaëtan Hadjeres and his supervisors divided Bach's chorales up into 80 per cent which were used to train the algorithm and 20 per cent that would be used as the test data. Volunteers were then asked to listen to chorales generated by DeepBach alongside real Bach chorales from the test data. They had to say whether they thought the chorale was by the computer or by Bach. Listeners were asked to report on their musical background, which will obviously impact the reliability of their assessment. (Composing students will hear things that an untrained ear might miss.)

The results were striking: 50 per cent of the time DeepBach pieces were believed to have been composed by Bach. Composition students had a slightly better hit rate, but even they had a hard time, failing to identify 45 per cent of DeepBach's compositions as fakes. This is impressive. Chorales are pretty unforgiving. It takes just one bum note to mark it out as a fake. Bach made no mistakes in his compositions, yet 25 per cent of his chorales were judged to have been cranked out by a machine! All very impressive. Without my wishing to sound snooty, chorales are possibly the dullest bit of Bach. Hymn tunes were something he needed to bash out, but they are not the Bach that really moves me.

One of the key difficulties with any project that tries to learn from the masters is a lack of good data. 389 chorales may seem like a lot, but it's really only barely enough to learn on. In successful machine-learning environments like computer vision, the algorithm will train itself on millions of images. Here it has only 389 data points, and most composers were far less prolific. Bach's chorales are useful in that they offer very similar examples of a single phenomenon. But when you look more broadly at a composer's output, there can be so much variety that a machine will be lost trying to learn from it. Maybe this is what will ultimately protect human-generated art from the advance of the machines. The good stuff is just too small in number for machines to learn how to replicate it. Sure they can churn out muzak, but not quality music.

12

THE SONGWRITING FORMULA

Music expresses that which cannot be put into
words and that which cannot remain silent.

Victor Hugo

I'm a trumpeter but I've never been able to master improvisational jazz. I have no problem playing sheet music in the orchestra, but jazz demands that I become a composer. Not just that, but a composer composing on the fly, responding in real time to the musicians around me. I have always had the greatest admiration for those who can do that.

Various attempts at learning jazz have taught me that there is a puzzle element to a good improvisation. Generally a jazz standard has a set of chords that change over the course of a piece. The task of the trumpeter is to trace a line that fits the chords as they change. But your choice also has to make sense from note to note, so playing jazz is really like tracing a line through a two-dimensional maze. The chords determine the permissible moves vertically, and what you've just played determines the moves horizontally. As jazz gets freer, the actual chord progressions become more fluid and you have to be sensitive to your pianist's possible next move, which will again be determined by the chords played to date. A good improviser listens and knows where the pianist is likely to head next.

To create a machine that can do this does not seem impossible, but there are challenges to overcome that algorithmic composers like Emmy don't face. A jazz-improvising algorithm has to play and respond to new material in a real-time interaction.

One of the classic texts on which many young musicians cut their teeth is *The Jazz Theory Book* by Mark Levine, who played alongside Dizzy Gillespie and Freddie Hubbard, two of the greatest jazz improvisers of the last century. As Levine points out: 'A great jazz solo consists of 1% magic and 99% stuff that is Explainable, Analyzable, Categorizeable, Doable.' That's all stuff that can be put into an algorithm.

Miles Davis's *Kind of Blue* is my favourite jazz album of all time. So how close are we to creating a Kind of DeepBlue?

Pushkin, poetry and probabilities

As a young man, François Pachet fantasised about being a musician, composing hits and playing the guitar like his heroes, but, despite some fair attempts at writing music, he was eventually seduced into a career in AI. While the head of the Sony Computer Science Laboratory in Paris, Pachet discovered that the tools he was learning in AI could help him compose music. He created the first AI jazz improviser using a mathematical formula from probability theory known as the Markov chain.

Markov chains have been bubbling under many of the algorithms we have been considering thus far. They are fundamental tools for a slew of applications: from modelling chemical processes and economic trends to navigating the internet and assessing population dynamics. Intriguingly, the Russian mathematician Andrey Markov chose to test his theory not on science but on Pushkin's poetry.

Markov's discovery emerged out of a dispute with another Russian mathematician, Pavel Nekrasov. One of the central pillars of probability theory is the law of large numbers, which states that if you have a coin and each toss of the coin is totally independent from the previous toss, as you toss the coin more frequently the number of heads and tails will get closer and closer to a fifty–fifty split. After four tosses there is a 1 in 16 chance that all tosses will be heads. But as we increase the number of tosses, the chances of deviating from fifty–fifty decrease.

Nekrasov believed the inverse was true: that if statistics followed the law of large numbers, actions must also be independent of previous results. He tried to use this to prove that because crime statistics in Russia obeyed the law of large numbers, criminals must be acting freely in their decision to commit a crime.

Markov was dismayed by Nekrasov's faulty logic. He described his work as 'an abuse of mathematics' and was determined to prove him wrong. He needed to cook up a model where the probability of an outcome depended on previous events and yet still the long-term behaviour obeyed the law of large numbers. Tossing a coin does not depend on previous tosses, so it was not what Markov was after. But what about adding a little dependence so that the next event depends on what just happened, but not on how the system arrived at the current event. A series of events where the probability of each event depends only on the previous event became known as a Markov chain. Predicting the weather is a possible example. Tomorrow's weather is certainly dependent on today's but not particularly dependent on what happened last week.

Consider the following model. It can be sunny, cloudy or rainy. If it is sunny today there is a 60 per cent chance of sun tomorrow, a 30 per cent chance of clouds and a 10 per cent chance of rain. But if it were to be cloudy today the probabilities would be different. Now there would be a 50 per cent chance of rain tomorrow, a 30 per cent chance it would remain cloudy and a 20 per cent chance of sun. In this model the weather tomorrow only depends on the weather today. It doesn't matter if we've had two weeks of sun – if it's cloudy today, the model will still give us a 50 per cent chance of rain tomorrow. The last part of the model tells us the probability of going from a rainy day today: we have 40 per cent chance of a sunny day, 10 per cent chance of a cloudy day and 50 per cent chance of another rainy day. Let us record these probabilities in what we call a matrix.

$$
\begin{pmatrix} SS & SC & SR \\ CS & CC & CR \\ RS & RC & RR \end{pmatrix} = \begin{pmatrix} 0.6 & 0.3 & 0.1 \\ 0.2 & 0.3 & 0.5 \\ 0.4 & 0.1 & 0.5 \end{pmatrix}
$$

With this model we can calculate what the probability of rain will be in two days' time from a sunny day. There are several ways to get there, of course, so we need to sum up all of the possible probabilities. It could go SSR (sunny, sunny, rain), SCR (sunny, cloudy, rain) or SRR (sunny, rain, rain).

the probability of SSR = probability of SS x probability of SR = 0.6 x 0.1 = 0.06

the probability of SCR = probability of SC x probability of CR = 0.3 x 0.5 = 0.15

the probability of SRR = probability of SR x probability of RR = 0.1 x 0.5 = 0.05

This means that the probability of rain two days from a sunny day, which we'll denote as SxS = 0.06 + 0.15 + 0.05, is 0.26 or a 26 per cent chance.

There is a convenient tool for calculating the chance of rain on the second day. It involves multiplying two copies of our probability matrix together.

$$\begin{pmatrix} SxS & SxC & SxR \\ CxS & CxC & CxR \\ RxS & RxC & RxR \end{pmatrix} = \begin{pmatrix} 0.6 & 0.3 & 0.1 \\ 0.2 & 0.3 & 0.5 \\ 0.4 & 0.1 & 0.5 \end{pmatrix}^2$$

Despite this dependence from day to day on the previous day's weather, in the long run, whether we start on a sunny, rainy or cloudy day, the chance of rain will tend towards the same value (about 32.35 per cent). To see this, we can multiply together more and more of our probability matrices and we will find that the entries on each row tend towards the same number. The long-term weather forecast is thus independent of today's weather, even if tomorrow's weather is dependent on it.

$$\begin{pmatrix} 0.6 & 0.3 & 0.1 \\ 0.2 & 0.3 & 0.5 \\ 0.4 & 0.1 & 0.5 \end{pmatrix}^{10} = \begin{pmatrix} 0.4412 & 0.2353 & 0.3235 \\ 0.4412 & 0.2353 & 0.3235 \\ 0.4412 & 0.2353 & 0.3235 \end{pmatrix}$$

Each row of this matrix represents the chance of a sunny, cloudy or rainy day after ten days. We can see now that it doesn't matter what today's weather is (i.e. which row we choose): the probability on the tenth day will always be the same. Markov had devised a proof that showed conclusively that Nekrasov's belief that long-term crime statistics implied the existence of free will was flawed.

Markov decided to illustrate his model with the help of one of Russia's most cherished poems, Pushkin's *Eugene Onegin*. He had no hope of providing new literary insights into the poem: he simply wanted to use it as a data set to analyse the occurrence of vowels and consonants. He took the first 20,000 letters, about an eighth of the poem, and counted the number of vowels and consonants. A computer could have done this in a flash, but Markov sat down and counted the letters by hand. He eventually concluded that 43 per cent were vowels and 57 per cent consonants. If you were to take a letter at random, you would therefore be better off guessing that it would be a consonant. What he was interested in figuring out was whether knowing the previous letter would change your guess. In other words, does the chance that the next letter will be a consonant depend on whether the previous letter is a consonant?

After analysing the text, Markov found that 34 per cent of the time a consonant would be followed by another consonant, while 66 per cent of the time it would be followed by a vowel. Knowledge of the previous letter thus changed the chances of a given outcome. This is not unexpected: most words tend to alternate between consonants and vowels. The chance of a vowel following a vowel, he calculated, was only 13 per cent. *Eugene Onegin* therefore provided a perfect example of a Markov chain to help him explain his ideas.

Models of this sort are sometimes called models with amnesia: they forget what has happened and depend on the present to make their predictions. Sometimes the model may be improved by considering how the two previous states might affect the next state. (Knowledge of the two previous letters in Pushkin's poems might help sharpen your chances of guessing the next letter.) But at some point this dependence disappears.

'The Continuator': the first AI jazz improviser

Pachet decided to replace Pushkin with Parker. His idea was to take the riffs of a jazz musician and, given a note, to analyse the probability of the next note. Let's imagine a riff consisting of an ascending and descending scale. If you play a particular note, the chances are 50–50 that the next note will be one up or one down the scale. Based on this fact, the algorithm would do a random walk up and down the scale. The more riffs you gave it, the more data it would have to analyse and the more a particular style of playing would emerge. Pachet figured out that it wasn't enough to look one note back, and it might take a few notes to know where to go next. But you don't want the algorithm to reproduce the training data, so it's no good going too far back.

The advantage of Pachet's approach is that the data can be fed in live. You could riff away on the piano. The algorithm would statistically analyse what you were up to and the moment you stopped, it would take over and continue to play in the same style. This form of question and response is common in jazz, so the algorithm could play with a live musician handing the melody back and forth. The algorithm became known as 'The Continuator', as it continued in the style of the person feeding it training data.

After each note The Continuator calculates where to go next based on what it has just played and what the training data tells it are the probabilities of hitting certain notes. Then it rolls the dice and makes a choice. In another version of the algorithm, which Pachet calls the 'Collaborator Mode' (rather than 'Question and Answer'), a human plays a melody and The Continuator uses its calculus of probabilities to guess at the right chord to play, much as a human accompanist would.

What did the jazz musicians who played with the algorithm think of the result? Bernard Lubat, a contemporary jazz musician who tested The Continuator out, admits to being quite impressed: 'The system shows me ideas I could have developed, but that would have taken me years to actually develop. It is years ahead of me, yet everything it plays is unquestionably me.' The Continuator had learned to master Lubat's sound world, but rather than simply throwing stuff back that he had done before, it was exploring new territory. Here was an algorithm that was demonstrating exploratory creativity. Beyond this, it was pushing the artist on whose work it had trained to be more creative by showing him aspects of his craft he had not accessed before.

For me this is the moment when the Lovelace Test was passed. It is the musical version of move 37 in game 2 of AlphaGo's contest with Lee Sedol. The algorithm is producing a result that is surprising both the programmers of the algorithm and the musician whom it trained on. And it isn't just new and surprising. For Lubat, the algorithm was helping him to be more creative. Its output was extraordinarily valuable, changing the way Lubat approached his practice.

We all tend to get stuck in our ways. The Continuator was initiating new sound sequences and effectively saying: 'Hey, you know you can do this too?' 'Because the system plays things at the border of the humanly possible,' Lubat explained, 'especially with the long but catchy melodic phrases played with an incredible tempo, the very notion of virtuosity is challenged.'

Lubat felt he was physically constrained in a way that The Continuator was not, and that this made it possible for The Continuator to be more innovative than he had been. Often lack of embodiment hinders computer creativity, but in this case we see the opposite. The fact that machines can do things so much faster and process so much more data than humans may result in an interesting tension between human creativity and AI creativity. This is the dynamic suggested by *Her*, a movie about a man who falls in love with his AI. After many hours talking together, the AI complains about how slow interactions are with humans and she ultimately leaves her human lover for more rewarding relationships with other AI that can interact at the speed of her CPU. Maybe The Continuator will start to produce sounds that only another machine will be able to appreciate because of its complexity and speed.

In the meantime, The Continuator elicited interesting emotional responses from its audiences. In live performances with Lubat jamming alongside it, Pachet reports that 'audience reactions were amazement, astonishment, and often a compulsion to play with the system'. Pachet decided to put The Continuator through a jazz version of the Turing Test. He got two jazz critics to listen to the jazz pianist Albert van Veenendaal improvising in a call-and-response mode with The Continuator. Both critics found it very difficult to distinguish one from the other, and thought The Continuator was likely to be the human jazz musician, as it was pushing the limits of the genre in more interesting ways.

The Continuator has broken down boundaries and done remarkable things, but systems based on Markov chains have certain inbuilt limitations. Although it produced musical riffs that locally made sense and were even quite surprising, its compositions were ultimately unsatisfying because they didn't have global structure or what we might call composition. Pachet realised he would have to constrain the evolution of the melody if it was going to have a more interesting story to tell. In question and response you often want the response to end where the question started, but you want the melody ultimately to realise some resolution of tension. To do this within the parameters of the Markov model was going to be like squaring a circle. Pachet decided he would have to find a new way to combine the freedom of the Markov process with the constraints that would lead to a more structured composition.

The Flow Machine

Many artists and performers claim that when they are totally engaged in their art they lose all sense of time and place. Some call this 'being in the zone'. More recently it has come to be known as 'flow', a term first identified as a psychological state of mind by the Hungarian psychologist Mihaly Csikszentmihalyi in 1990. Pachet decided he would try to create an algorithm that would help get creative artists into a state of flow.

Flow is achieved at the meeting point of extreme skill and great challenge. Without either one of these two you slip into one of the other psychological states identified in the diagram on page 222. If you don't have the skills and you try something too

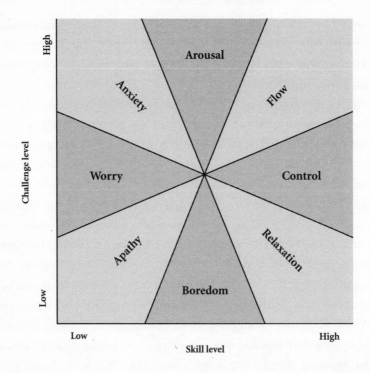

challenging, then you end up in a state of anxiety. If something is too easy for you given your skill set, then you are likely to border on boredom.

The algorithm at the heart of Pachet's Flow Machine uses the Markov processes to learn the style of an artist and then provides certain constraints. This is how many creative artists work. Picasso spent years absorbing the work of El Greco, Renoir, Velázquez and Manet, imitating, combining and adapting their styles and passing them through different sets of constraints he imposed to arrive at a style that was uniquely his own and yet rooted in the masters of the past.

Pachet experimented with getting his algorithm to play in one style while taking constraints from another. This was a fantastic example of algorithmic experimentation with Boden's concept of combinational creativity. In one instance he took Charlie Parker's style of the blues and combined it with the constraints offered by the serial world of Pierre Boulez, who, as a fan of Schoenberg, made sure his pieces used all twelve notes of the chromatic scale. This forces the blues to cycle through all twelve notes, an unusual constraint as the blues will usually only use three notes as a foundation for its progression. The result is a strange beast that clearly emerged from the bebop world of Charlie Parker, but lives well on the fringes of that world. I must admit to enjoying the fusion. Others didn't work so well, like John Coltrane playing 'Giant Steps' with the chord constraints of Richard Wagner!

The Flow Machine isn't limited to music. You can get it to learn the style of one poet or lyricist and apply it to the constraints of another. For example, Pachet's team got their Markov models to capture the style of Bob Dylan's lyrics and then applied it to the lyrics of 'Yesterday' by the Beatles. The Beatles' lyrics provided certain constraints of metre and rhythm. The Flow Machine was then tasked with filling in this framework with a choice of words and phrases that would be recognisably Dylan. Here is the result, which you can sing to the tune of 'Yesterday':

> Innocence of a story I could leave today
> When I go down on my hands and pray
> She knocked upon it anyway
> Paradise in the dark side of love it is a sin
> And I am getting weary looking in
> Their promises of paradise
> Now I want to know you would be spared this day
> Wind is blowing in the light in your alleyway
> Innocence in the wind it whispers to the day
> Out the door but I could leave today
> She knocked upon it anyway

Pachet used his Flow Machine to create what has been trumpeted as the first pop song written by AI. Finally his years of research had allowed him to realise his childhood dream. The new song for which the Flow Machine composed the music is called 'Daddy's Car', and is written in the style of Pachet's favourite band: the Beatles. Many musical analysts have argued that there is a secret formula to the music of the Beatles and Pachet hoped to crack their code. But his lyrics were not actually generated by algorithms. They were the work of Benoît Carré, who was also given the job of turning the output of the algorithm into a fully produced track.

'Daddy's Car' was followed by the album *Hello World,* released early in 2018. The title is a reference to the first exercise anyone learning to code is made to do: create a program that outputs the text 'Hello World'. The album is a collaboration between Carré and other musicians who have used the Flow Machine to push the boundaries of their own creativity. To say that it is the first album produced by AI is not quite accurate, as Carré and his collaborators played an important part in determining the contours of the final product.

The result? The composer Fatima Al Qadiri dismissively quipped that 'it sounds like a song that has been Xeroxed fifty times and then played'.

But not everyone was so negative. Pachet has been poached from Sony Labs and is now working for Spotify. Given that rumours have been circulating that Spotify is creating playlists full of songs by 'fake' artists, the move is an interesting one. Music critics spotted a number of artists on Spotify who were notching up an extraordinary number of hits thanks to their inclusion on popular playlists curated by Spotify for meditation or running. A band called Deep Watch had recorded 4.5 million plays over a five-month period.

When critics tried to find out who these artists were, they kept drawing blanks: no presence anywhere else on the internet; no upcoming concerts; no details anywhere of such a band. A rumour began to circulate that the music was being generated by 'fake artists' so that Spotify wouldn't have to pay royalties.

Spotify hit back: 'We do not and never have created "fake" artists and put them on Spotify playlists. Categorically untrue, full stop.' But it does appear that they were specifically commissioning minor artists to create songs under fake names at royalty rates that were much more favourable to the company than its standard deals with record labels.

The fact that it was possible for these composers to knock out an endless supply of mediocre pop songs is a result of the incredibly formulaic nature of the genre. Unlike the subtleties of many classical composers, so many pop songs just replicate tried-and-tested formats without challenging expectations. Most have four beats to the bar, the tune comes in chunks of four or eight bars, with melodic fragments that are repeated over and over so that people can quickly sing along, and the song never changes key. Of course, there are exciting moments when songs break the rules, but often they just create a new formula that then gets repeated over and over.

Will the hiring of Pachet by Spotify up the game and result in even these artists being put out of a job? Algorithms are already curating our listening. How long will it be before the songs we listen to are created for us bespoke by algorithms? Spotify will no longer need to pay any royalties at all – just Pachet's salary.

If you want your very own piece of AI-generated music, you can visit Jukedeck's website. Set up by two Cambridge graduates who met as choir boys at the age of eight, Jukedeck is one of a number of companies using AI to generate songs for organisations, from the Natural History Museum to Coca-Cola. These companies need original but cheap background tracks for videos and promotional material. It doesn't have to be the best track ever. They don't want to have to pay exorbitant royalties. The tracks generated by Jukedeck are perfect aural filler for their video content.

The website allows you to choose different genres of music from folk to chill-out, from corporate (is that a genre?) to drum and bass. Then you can tell it if you want your music to be aggressive or melancholy or any one of eight other moods. Once you've chosen and clicked, the algorithm churns out ninety seconds of music and even names the track for you.

My choice of cinematic sci-fi music produced a track called 'Impossible Doubts'. It isn't something I'll be listening to regularly, but that's not the point. The phrase 'good enough' is one that is bandied around a lot in the AI music generation scene. Jukedeck is aiming at the market for background music for video production or game development, not to compete with Adele. With an algorithm that can react to mood it is a perfect tool to follow the trajectory of a player as they navigate a game. If I want to listen to 'Impossible Doubts' I can get a royalty-free licence to use it for $0.99 or I can buy the copyright and own the track outright for $199.

Perhaps those dollar signs are an important indicator of the drive behind AI-generated music. It is less artistic considerations that are driving the AI art revolution than money.

Quantum composition

One of the curious aspects of artistic creation is the idea that an artist creates one piece that has to appeal to the many different people who will view, read or listen to it. But each listener has different tastes, expectations and moods. What if you could create art that reversed this idea of one for many and sought instead to go from many works to one individual? Our smartphones gather a huge amount of data about us. What if all of that information could be used to create a bespoke piece of art just for you?

This is exactly what Massive Attack decided it would do. After releasing nothing new since their album *Heligoland* in 2010, the band chose an innovative way to release four new songs at the beginning of 2016. Fans could only access the songs by down-loading a new purpose-built app called Fantom. Then came the interesting twist: once you had allowed the app to access your location, time of day, camera view, heartbeat and Twitter feed, an algorithm would decide how to mix the tracks for you.

Massive Attack's algorithm was essentially playing a more sophisticated version of Mozart's game of dice. The original track was broken down into small mini-tracks that could be used as the raw ingredients for the creation of new personalised tracks. At each point in the evolution of the new song, choices are made that determine which micro-track will be added next and how it will be mixed. These decisions are guided by the data the algorithm gathers from its individual users. If your heart rate is up and you are moving fast and the camera is picking up bright colours, this information will influence the tone and texture of the song you hear.

The art lies in creating a tree of possibilities that will be sufficiently rich and varied but also sufficiently coherent, so that whatever path the algorithm takes, the result appears seamless and natural. What you don't want is total randomness. Mozart carefully curated each bar, offering eleven options, each one of which might make sense as the next bar in the waltz. The overall structure of the waltz set up the rules within which the game could be played. The same is true of Massive Attack's algorithm. You don't want the chorus crashing in during the evolution of the verse.

Rob Thomas, the programmer who helped create the app, rather nicely refers to the result as 'quantum composing'. In the quantum world, an electron can be in many different places at the same time thanks to something called quantum superposition. It is the act of observation that forces the electron to collapse into one of its many possible states. The idea, for Thomas, was to create a song that could exist in many possible states. When I decide to listen to the song, the algorithm takes my data and chooses how to collapse Massive Attack's 'wave function' into a single song for me.

Thomas is interested in the dialogue between our emotional states and the music we listen to, and how one can influence the other. 'Music is this emotional manipulation tool,' he says. 'I want to know how I can use these musical tactics to induce an emotional state in the people listening.' He is currently exploring the use of AI music to help induce a meditative state in apps dedicated to mindfulness. The idea is that the music reacts to data about the current state of your mind and body in the hope of learning how to manipulate your body into relaxing. Of course, if you want to harness the most effective emotional manipulator, Thomas admits, you really need to create a human being. He laughs as he concedes: 'There are much easier and more pleasurable ways to make humans than using AI.'

The Fantom app depends on the musician's ability to curate the component pieces of a song. But Massive Attack recognises the power of machine learning to create its tree structure of possible choices in a much more organic manner. The band hopes, with its next release, to let machine learning create its own versions of tracks. Rob Thomas has teamed up with Mick Grierson at Goldsmiths, University of London, to realise this next step.

Grierson has worked closely with the Icelandic avant-garde band Sigur Rós. He took one of their songs, 'Óveður', and extended it into a twenty-four-hour version that never repeats itself and yet retains the sound of the original five-minute track. The twenty-four-hour version was created to accompany a journey around the coast of Iceland that was televised and broadcast on YouTube and Iceland's National TV. Part of the new craze for Slow TV, the journey began on the eve of the summer solstice on 20 June 2016. The artists travelled 1332 kilometres anticlockwise around Iceland's coastal route 1, passing by Vatnajökull, Europe's largest ice-sheet, the glacial lagoon, the East Fjords and the desolate black sands of Möðrudalur.

For a human composer to create a twenty-four-hour sound-track that doesn't repeat itself would be tough and expensive. The software developed by Grierson uses probabilistic tools to gener-ate the track in response to whatever images it is accompanying. He's since created a longer version of the song that will play for ever, never repeating itself. When Massive Attack and Sigur Rós disband, these algorithms will make it possible for you to keep listening to new versions of their songs for as long as you want.

Brian Eno coined the phrase 'generative music' to describe music that is ever-changing, created by a system or algorithm. Eno likes to say it is music that thinks for itself. It's a sort of musical garden, where the composer plants the seeds and the interaction of the algorithm with the outside world – a human playing a computer game or how the listener makes their way through the day – grows the sound garden from these seeds. In some ways live performances are meant to capture this idea that the jour-ney from score to experience will produce something unique each time. Eno was interested in pushing this idea further. His apps like Bloom or Scape, or his most recent one, Reflection, created with Peter Chilvers, produce endless Eno-like music that is gener-ated by users interacting with the screen on their smartphones. He describes the process of generation as being like watching a river: 'it's always the same river, but it's always changing'.

Eno has embraced technology in his creations, but like Love-lace he does not believe the algorithms he works with will ever generate anything more than what their creators put in. 'There's quite a lot of intent in this and there have already been a lot of aesthetic choices made. When somebody gets this and makes a piece of music with it, they're making a piece of music in col-laboration with us.'

But machine learning is starting to challenge the Lovelace crutch that human composers are clinging to. In 2016 an algo-rithm called AIVA was the first machine to have been given the title of composer by the Société des auteurs, compositeurs et éditeurs de musique (SACEM), a French professional association in charge of artists' rights. Created by two brothers, Pierre and Vincent Barreau, the algorithm has combined machine learning with the scores of Bach, Beethoven, Mozart and beyond to produce an AI composer that is creating its own unique music. Although it's currently writing theme tunes for computer games, its aims are lofty: 'to make her mark in the timeless history of music'. Listening to AIVA's first album, called appropriately enough *Genesis*, I don't think Bach and Beethoven have much to worry about yet. But, as the title suggests, this is just the beginning of a musical AI revolution.

Why do we make music?

Music has always had an algorithmic quality to it, which means that of all the art forms it is most under threat from the advances of AI. It is the most abstract of all art forms, tapping into structure and pattern, and it is this abstract quality which gives it close ties to mathematics. But this means it inhabits a world in which an algorithm will feel as much at home as a human.

But music is more than just pattern and form. It has to be performed if it is going to be brought alive. Humans started making music to accompany certain rituals. Inside the caves whose walls our early ancestors daubed with paint, archaeologists have found evidence of musical instruments: flutes made from the bones of vultures; animal horns that could be blown like a trumpet; objects attached to strings that, when swung overhead, create a strange and eerie sound.

Some have speculated that these primitive instruments may have been used to communicate, but others believe they were an important component of the rituals our early ancestors began to develop. It seems that the need for rituals is very much a part of the human code. A ritual consists of a sequence of activities involving gestures, words and objects performed in a sacred place according to a set sequence or pattern. Often, from the outside, the ritual appears irrational or illogical, but for insiders it offers an important way of binding the group. Music plays an important part of many such rituals. Singing in a choir or playing in a band is an extraordinary way of uniting disparate conscious experiences. The songs we sing on the stands in sports games bind us as a crowd against the away fans.

That ability of music to bind a group may be what gave *Homo sapiens* an advantage when they migrated to Europe and encountered Neanderthals. As the composer Malcolm Arnold wrote: 'Music is the social act of communication among people, a gesture of friendship, the strongest there is.' The Paleolithic flutes dating back 40,000 years found in Germany may have allowed our ancestors to communicate with each other over large distances. It was quickly realised that music was a powerful ingredient in the creation of mind-altering rituals. Repetition can help alter our state of consciousness, as witnessed by many shamanic practices. Our brains have natural frequencies that correspond to different mental states. Trance music taps into the fact that a rhythm drummed out at 120 beats per minute is best tuned for inducing a hallucinatory experience in humans. We know from modern experiments that messing with multiple sensory input can cause the mind to have strange out-of-body experiences. The combination of touch and sight, for example, can cause someone to identify with a false limb. This is why together with those early instruments we often find spices or herbs that would have given the rituals a smell as well as a sound. How can an algorithm that is not embodied ever hope to understand the power of music to change our bodies and alter our minds?

As civilisations evolved, music continued to be part of the ritual world. The great advances in music from Palestrina to Bach to Mozart were often made in a religious context. There is some speculation that the concept of God emerged in humans with the emergence of our internal world. With the development of consciousness came the shock of being aware of a voice in your head. That must have been quite frightening. Ritual and music could appease that voice in the head, and the forces of nature that seemed to be a place for the gods.

This all sounds so far from the logical, emotionless world of the computer. Algorithms have certainly learned how to make the sounds that move us. Algoraves now use algorithms that react to the pulsating crowd to help DJs curate the sounds that will keep the crowds dancing. DeepBach is composing more religious chorales for church choirs to sing their praise to God. But despite the fact that these algorithms appear to have cracked the musical code, there is nothing stirring inside the machine. These are still our tools, the modern-day digital bullroarers.

13

DEEPMATHEMATICS

*It takes two to invent anything. The one
makes up combinations, the other one chooses.*

Paul Valéry

It was while sitting next to Demis Hassabis at one of the Royal
Society's meetings about the impact that machine learning was
going to have on society that I had an idea. It was Hassabis's
algorithm AlphaGo that had started my whole existential crisis
about whether the job of being a mathematician would continue
to be a human one. Hassabis and I had both recently been made
Fellows of the Royal Society, one of the highest accolades for a
scientist. So if Hassabis could get an algorithm to 9 dan status
in Go, could he get an algorithm to prove a mathematical theorem that might lead to it being elected as a Fellow of the Royal
Society?

But when I turned to Hassabis and threw down the gauntlet, I
got something of a surprise. 'We're already on the case,' he whis-
pered to me. It seems as if nothing has escaped their radar. As he
explained after the meeting had finished, he already had a team
in place trying to train algorithms to learn from the proofs of the
past to create the theorems of the future. Hassabis suggested I
drop by DeepMind's offices to find out how far they'd got.

It was with some trepidation that I set off to find out if

mathematics would soon be yet another casualty of the machine-learning revolution. Although DeepMind was purchased by Google in 2014 for £400,000,000, Hassabis had been determined that his baby should stay in London, and so the offices are part of Google's London campus just next to King's Cross. Walking through the station concourse, I spotted a long line of people hoping to have their picture taken next to Harry Potter's famous platform 9¾. It struck me that if they wanted to experience real magic, they should be heading next door.

The whole Google site has the feel of a modern Oxford college, conceived to provide the best environment for scholars to concentrate on deep thinking. Google employees are offered free food around the clock, while baristas are on hand to fuel their brains with caffeine. There are a 90-metre running track, free massages and even cookery classes by Dan Batten, a chef who worked with Jamie Oliver, although, given the free food, this seems to be more for entertainment than nourishment. And when their brains have gone into overdrive, Googlers can conk out in one of the Nap Pods dotted around the building.

This is all taking place at a temporary home while Google's cutting-edge new headquarters are going up next door. Designed by the Danish architect Bjarke Ingels and Thomas Heatherwick, the British designer behind London's 2012 Olympic cauldron, it promises to be an extraordinary building – some have referred to it as a 'landscraper' – as long as London's Shard is high. If other Google sites are anything to go by, it will be quite something. The building in Victoria has a room filled with musical instruments for employees to jam on during their downtime. The site in Mountain View, California, has its own bowling alley. The new campus at King's Cross will more than keep up with its rivals, with an Olympic-sized swimming pool and an amazing roof garden for employees to enjoy during breaks from coding, or even as a place to code if so inclined. The garden will be themed around three areas: 'plateau', 'gardens' and 'fields', planted with strawberries, gooseberries and sage. The opulence of the Google offices is a clear sign that the business of machine learning is booming. But for now I was heading for the tower block at number 6, Pancras Square.

DeepMind occupies two floors of the current campus. One is dedicated to commercial applications of their work, but it was to floor 6 that I was whisked, where research is being done. The programmers on floor 6 have an interesting range of projects in their crosshairs. Machine learning is being applied to help navigate the slippery random world of quantum physics, and projects are bubbling away to infiltrate biology and chemistry. But I was interested in their work on maths.

Hassabis had suggested I speak to Oriol Vinyals to find out how far along they were in their effort to generate an original mathematical proof. Originally from Spain, where he studied mathematics as an undergraduate, Vinyals soon knew that his passion was for artificial intelligence. So he headed to California to do his doctorate, where he was picked up by Google Brain and then DeepMind.

I must admit to being both concerned and excited as the door of the lift opened and Vinyals greeted me. But I very quickly felt at ease. Like many of the people wandering around the Google campus, Vinyals would easily fit into my department in Oxford. This was not a corporate environment but a place where T-shirt and jeans are acceptable wear (provided your T-shirt bears some suitable nerdy caption).

We made our way to one of the meeting rooms, all of which were named after scientific pioneers like Ada Lovelace, the room, appropriately enough, that we found ourselves in. Vinyals explained that it wasn't just researchers at DeepMind who were involved but also Google researchers dotted around the world. So what sort of mathematics were these Googlers exploring? Had they chosen to tackle a theorem in my own world of symmetry? Or to prove something about networks and combinatorics? Or to determine whether variants of Fermat's equations have solutions? Vinyals soon revealed that they were going for a very different angle to the one I had expected, one that felt quite alien to what I think mathematics is about.

The mathematics of Mizar

The team at DeepMind and Google had decided to focus on a project that began in Poland in the 1970s called Mizar. The aim of the project was to build up a library of proofs that were written in a formal language which a computer could understand and check. The brain behind Mizar was the Polish mathematician Andrzej Trybulec, but it was his wife who was responsible for the name. She'd been looking through an astronomical atlas when her husband had asked her for a good name for his project and she had suggested Mizar, a star in the Big Bear constellation.

Anyone was open to submit a proof written in this formal language, and by the time Trybulec died in 2013 the Mizar Mathematical Library had the largest number of computerised proofs in the world. Some of the proofs were constructed by humans but written in this computer language, while others had been generated by the computer. The project is currently maintained and developed by research groups at Białystok University in Poland, the University of Alberta in Canada and Shinshu University in Japan. Interest in the project had slowed in recent years and the library had not been growing fast. Little did they know that Deep-Mind and Google had set their sights on significantly expanding the library.

So far those who had been working on Mizar over the decades had successfully created a database containing more than 50,000 theorems. Given that the proofs in the database are written in a language that a computer can understand rather than a human, those involved in the Mizar project have been keen to pick out the theorems that human mathematicians would recognise as some of their all-time favourites. For example, there is a formalised computer proof of the Fundamental Theorem of Algebra, which states that every polynomial of degree n has n solutions in the complex numbers.

It was interesting to see this theorem in there. The human journey went through so many different false proofs, starting in the early seventeenth century and including false proofs by such eminent mathematicians as Euler, Gauss and Laplace. The first proof to be recognised as complete was finally made by Jean-Robert Argand in 1806. The gaps in the previous proofs were often quite subtle. It took time for the mistakes to be spotted. But once a proof that a computer could check had been found, there was a great confidence in its validity.

The way a computer generates a proof to include in the Mizar Library shares something in common with playing a game. You start with a list of basic axioms about numbers and geometry. You are allowed certain rules of inference. And from there it maps out pathways to new statements that are linked together by a string of rules of inference. It is, in some sense, similar to playing the game of Go. The board starts empty. The rules of inference are that you are allowed to place a stone (which must alternate between black and white) on the board in any position not previously occupied by a stone. The theorems are like endgames that you are trying to reach.

This is what the DeepMind team realised. Proving theorems and playing Go are conceptually related: both consist of searching for specific points in a tree of possible outcomes. Each point could branch off in many different directions and the length of each branch before it reaches its endpoints could be extremely long. The challenge is how to evaluate which direction one should head off in next to reach a valuable endpoint: to win the game or prove a theorem.

This model suggests you can unleash a computer and start

generating theorems. But this is not so interesting. There would be lots of overlap as you could reach the same endpoint in multiple ways. The real question was whether, given a statement or potential endpoint, you could find a pathway to that statement, a proof. If not, was there a pathway to prove its negation?

When the team at DeepMind and Google started looking at the theorems on Mizar's books, it found that 56 per cent had proofs which had required no human involvement. Their aim was to increase this percentage. The object was to create a new theorem-proving algorithm that would use machine learning to train on those proofs that had been successfully generated by the computer. The hope was that the algorithm could learn good strategies for navigating the tree of proofs from the data already in the Mizar Mathematical Library. In the paper that Vinyals proudly handed to me, the DeepMind and Google team had upped the percentage of computer-generated proofs in the library using their proof-generating algorithm from 56 per cent to 59 per cent. Although that may not sound remarkable, you have to recognise that this represents a non-trivial step change by applying these new techniques. This isn't just one extra theorem or one new game won. This is a 3 per cent increase in proofs that computers can reach.

In some ways I could see why Vinyals was excited by the progress. It was like an algorithm learning to play jazz, but instead of deciding the best choice of note to play next, it was deciding which logical move to make. The algorithm had expanded the reach of the computer in a significant way. It had pushed into new territory. Like new music, the computer had generated new theorems.

Yet I must admit I left the DeepMind offices rather downcast. I should have been elated by such a surge in mathematical progress, but what I had seen was like a mindless machine cranking out mathematical muzak, not the music of the spheres that gets me excited. There was no judgement being made about the value of these new discoveries, no interest in whether any of them contained surprising revelations. They were just new. They seemed to be missing two-thirds of what makes a creative act.

A mathematical Turing Test

Was this really the future? I went back and tried to read some of the proofs in Mizar's Library of my favourite theorems. They left me cold. Actually, they left me confused because they didn't speak to me at all. I could barely navigate their impenetrable formal language. I experienced what most people probably feel when they open one of my papers and see a string of seemingly meaningless symbols. The proofs were written in computer code that allowed the algorithms to formally move from one true statement to the next. This was fine for the computer, but it's not how humans communicate mathematics. For example, here is Mizar's proof that there are infinitely many primes:

```
reserve n,p for Nat;
theorem Euclid: ex p st p is prime & p > n proof
set k = n! + 1;
n! > 0 by NEWTON:23;
then n! >= 0 + 1 by NAT1:38; then k >= 1 + 1 by REAL1:55;
then consider p such that
A1: p is prime & p divides k by INT2:48; A2: p <> 0 & p > 1
    by A1,INT2:def 5; take p;
thus p is prime by A1;
assume p <= n;
then p divides n! by A2,NATLAT:16;
then p divides 1 by A1,NAT1:57;
hence contradiction by A2,NAT1:54;
end;
theorem p: p is prime is infinite
from Unbounded(Euclid);
```

Totally impenetrable even for a professional mathematician like me! It in no way corresponded to the way any human would ever tell the story. The problem, at some level, was a language barrier.

If algorithms can be written to translate Spanish to English, is there a way to translate from computer speak to the way a human would communicate a proof? This was a proposition that two Cambridge mathematicians, Timothy Gowers and Mohan Ganesalingam, set out to explore. Gowers first hit the headlines when he won a Fields Medal in 1998 and was then elected Rouse Ball Professor the same year.

Ganesalingam began by following a similar trajectory, studying mathematics at Trinity College, Cambridge, but then, after being selected as Senior Wrangler and receiving one of the top degrees in his year, he decided to shift paths and surprised everyone in his department by getting a Masters degree in Anglo-Saxon English. He won a university prize for the best results in the Cambridge English Faculty that year and went on to do a PhD in computer science, analysing mathematical language from a formal linguistic point of view. This combination of mathematics and linguistics would soon be put to use. Gowers and Ganesalingam's paths crossed in Trinity and they soon realised they were both interested in the challenge of the impenetrable nature of computer language. They decided to team up to create a tool to produce computer proofs that could be read by humans.

To test how good their algorithm was they tried an experiment on Gowers' blog. Gowers presented five theorems about metric spaces, a subject taught to first-year undergraduates, and included three proofs of each theorem. One was written by a PhD student, one by an undergraduate and one by their algorithm. So as not to prejudice the results, readers of the blog were not told the origin of the proofs. Gowers simply asked people to provide their opinions on the proofs' quality. They were asked to grade each one. He wanted to see if anybody would suspect that not all the write-ups were human-generated. Not one person gave the slightest hint that they did. In a second blog post he then revealed that one of the proofs had been written by a computer. Now he asked his participants to try to identify the computer proof.

The computer was typically identified by around 50 per cent of all those who voted. Half of these were confident that they were correct, and half not so confident. A non-negligible percentage of respondents claimed to be sure that a write-up that was not by the computer was by the computer. It was generally the undergraduate's answer that was wrongly believed to have been produced by a computer.

So how does a Fields Medal winner feel about computers muscling in on his patch? In his blog Gowers writes:

> I don't see any in-principle barriers to computers
> eventually putting us out of work. That would be sad,
> but the route to it could be very exciting as the human
> interaction gets less and less and the 'boring' parts of the
> proofs that computers can handle get more and more
> advanced, freeing us up to think about the interesting
> parts.

But it wasn't just the linguistic problem of the Mizar project that was bugging me. Of those extra 3 per cent of theorems that the DeepMind and Google team had managed to generate, were there any that would surprise me or make me gasp? I began to feel that this whole project missed the point of doing mathematics. But what exactly is the point?

The mathematical Library of Babel

One of my favourite short stories will help me answer that question. 'The Library of Babel', by Jorge Luis Borges, tells the story of a librarian's quest to navigate the contours of his library. He begins with a description of his place of work: 'The universe (which others call the library) is composed of an indefinite and perhaps infinite number of hexagonal galleries . . . From any of the hexagons one can see, interminably, the upper and lower floors.' There is nothing other than the library. This library of course is a metaphor for our own library (which we call the universe). As befits a library, this vast beehive of rooms is full of books. The tomes all have the same dimensions. Each one is 410 pages, each page has forty lines and each line consists of eighty orthographical symbols of which there are twenty-five in number.

As the librarian explores the contents of his library he finds that most of the books are formless and chaotic in nature, but every now and again something interesting appears. He discovers a book with the letters 'MCV' repeated from the first line to the last. In another, the cacophony of letters is interrupted on the penultimate page by the phrase 'Oh time thy pyramids' and then continues its meaningless noise.

The challenge the librarian sets himself is to determine whether the library is in fact infinite, or, if not, what shape it is. As the story develops, a hypothesis about the library is proposed: 'The Library is total . . . its shelves register all the possible combinations of the twenty-odd orthographical symbols (a number which, though extremely large, is not infinite): in other words, all that it is given to express, in all languages. Everything.' The library contains every book that it is possible to write. Tolstoy's *War and Peace* is somewhere to be found on the shelves. Darwin's *On the Origin of Species*. Tolkien's *The Lord of the Rings*, together with translations of all these works into all languages. Even this book is somewhere among the tomes shelved in the library. (At this point, having only got this far in my writing, how I would dearly love to find that book and spare myself the labour of carving out the rest of the text!)

Given that all the books have the same dimensions, it is possible to count how many books there are. If there are twenty-five symbols (which presumably includes spaces, full stops and commas) then I have twenty-five choices for the first character and twenty-five choices for the second character. That's already $25 \times 25 = 25^2$ possible choices just for the first two characters. There are eighty characters in the first line. Given twenty-five choices for each place, that gives 25^{80} possible first lines.

Now expand this to count the number of possible first pages. We get $(25^{80})^{40} = 25^{80 \times 40}$ different possibilities since there are forty lines on each page. Now we can get the total number of books in the library. This comes to $(25^{80 \times 40})^{410} = 25^{80 \times 40 \times 410}$ possible books. This is a lot of books. Given that there are only 10^{80} atoms in the observable universe, even if each atom was a book we wouldn't get anywhere near the total number of books in the Library of Babel. But it is still a finite number. We could easily program a computer to systematically generate all the books in a finite amount of time. Admittedly, the current estimate for the time we've got till the universe decays into a cold dark place means that practically this wouldn't be possible, but let's stay in the realm of theory and story.

When it was proclaimed that the library contained all books, the first impression was one of extravagant happiness. But this was followed by great depression, because it was realised that this library that contained everything in fact contained nothing. What makes my library, the Bodleian, which contains Tolstoy, Darwin, Tolkien, and will contain my book once it's published, different from the Library of Babel is the fact that a human being – or many human beings – has deemed this particular combination of letters worthy of its place in the Bodleian as part of our literary universe.

But what if we were to move to the mathematics section, which houses great journals like *Annals of Mathematics* and *Les Publications mathématiques de l'IHES*. What qualification is necessary to make it into one of the journals on those shelves? I think many people have the impression that this bit of the library aspires to be a mathematical Library of Babel, that the role of the mathematicians through the ages is to document all true statements about numbers and geometry. The irrationality of the square root of 2. A list of the finite simple groups. The formula for the volume of a sphere. The identification of the brachistochrone as the curve of fastest descent.

This is what Mizar was attempting to do. It has a list of mathematical statements and it is trying to see whether it can move from the opening axioms to these statements or their negation. The qualification for making it into the Mizar database is whether there is a proof of the statement. There are no choices being made based on what the statement means or whether someone thinks it is exciting enough to share with other mathematicians. It is simply a Library of Babel containing everything that it is possible to prove.

This, for me, goes against the spirit of mathematics. Mathematics is not a list of all the true statements we can discover about numbers. This may come as a shock to most non-mathematicians. Mathematicians, like Borges, are storytellers. Our characters are numbers and geometries. Our narratives are the proofs we create about these characters. And we make choices based on our emotional reaction to these narratives, deciding which ones are worth telling.

Let me quote one of my mathematical heroes, Henri Poincaré, explaining what it means to him to do mathematics: 'To create consists precisely in not making useless combinations. Creation is discernment, choice . . . The sterile combinations do not even present themselves to the mind of the creator.' Is mathematics created or discovered? The reason why we feel it is created comes down to that element of choice. Sure, someone else could have come up with it. But the same could be said of Eliot's *The Waste Land* or Beethoven's *Grosse Fuge*. There are so many different ways the notes could have been chosen that we can't imagine anyone else having composed these great works. The surprise, for most people, is that this same freedom exists in mathematics.

Mathematics, as Poincaré so beautifully put it, is about making choices. What then are the criteria for a piece of mathematics making it into the journals? Why is Fermat's Last Theorem regarded as one of the great mathematical opuses of the last century, while an equally complicated numerical calculation is seen as mundane and uninteresting? After all, what is so interesting about knowing that the equation $x^n + y^n = z^n$ has no whole number solutions when $n>2$?

This for me is where mathematics becomes more of a creative art than simply a useful science. It is the narrative of the proof of a theorem that elevates a true statement about numbers to the status of something deserving its place in the pantheon of mathematics. I believe a good proof has many things in common with a great story or a great composition that takes its listeners on a journey of transformation and change.

Mathematical fables

The best way to give you an idea of the narrative quality of a proof may be to tell you one of these mathematical stories. It is one of the first proofs I encountered when, at thirteen, I read G. H. Hardy's beautiful book *A Mathematician's Apology*. The book, which describes what it is like to be a mathematician, was heralded by Graham Greene as the best description of a creative artist since the diaries of Henry James.

Hardy presents in his book what is probably one of the first proofs in the history of mathematics, discovered by Euclid. The principal characters in our proof are prime numbers, numbers that are indivisible, like 3, 7 or 13. The narrative journey I want to take you on is to reveal that there are infinitely many of these characters and that if you were to try to list them all you would be writing for ever. I've already shown you the way Mizar tells the proof earlier in this chapter. Now let me tell you the story.

A proof is like the mathematician's travelogue. Euclid gazed out of his mathematical window and spotted this mathematical peak in the distance, the statement that there are infinitely many primes. The challenge for subsequent generations of mathematicians was to find a pathway leading from the familiar territory that mathematicians had already charted to this foreign new land.

Like Frodo's adventures in *The Lord of the Rings*, a proof is a description of the journey from the Shire to Mordor. Within the boundaries of the familiar land of the Shire are the axioms of mathematics, the self-evident truths about numbers, together with those propositions that have already been proven. This is the setting for the beginning of the quest. The journey from this home territory is constrained by the rules of mathematical deduction, which function like the legitimate moves of a chess piece, prescribing the steps you are permitted to take through this world. At times you will arrive at what appears to be an impasse and need to take a detour, moving sideways or even backwards to find a way around. Sometimes you need to wait for new mathematical characters like imaginary numbers or calculus to be created before you can continue your journey.

The proof is the story of the trek and a map charting the co-ordinates of that journey. It is the mathematician's log. A successful proof will function like a set of signposts to allow all subsequent mathematicians to make the same journey. Readers of the proof will experience the same exciting realisation as its author that this path allows them to reach that distant and seemingly impenetrable peak. Very often a proof will not seek to dot every *i* and cross every *t*, just as a story does not present every detail of a character's life. It is a description of the journey and not a re-enactment of every step. The arguments mathematicians provide are designed to guide the mind of the reader. Hardy described the arguments we give as 'gas, rhetorical flourishes designed to affect the psychology, pictures on the board in the lecture, devices to stimulate the imagination of pupils'.

It is a strange aspect of mathematical stories that they often begin with the ending. The challenge is to show how to reach this climax from our current state of the saga. The narrative journey requires some scene setting, mapping out the story so far and providing a brief description of familiar territory. We will be reminded that the important characteristic of prime numbers is that they are the building blocks of all other numbers. Every number can be built by multiplying prime numbers together: 105, for example, is constructed by multiplying 3 x 5 x 7. Or sometimes you need to repeat a prime; for example, 16 = 2 x 2 x 2 x 2.

So let us begin our journey to explain why there are infinitely many of these prime suspects. Suppose that this were not the case, and that we could make a list of these characters, a *dramatis personae*. This is a classical narrative device in the mathematician's toolbox. Like *Alice's Adventures in Wonderland* or *The Wizard of Oz*, imagine a world where the opposite of what you are trying to prove is true and let the narrative play out to its absurd conclusion.

Suppose, for a moment, that this *dramatis personae* consisted of the prime characters 2, 3, 5, 7, 11 and 13. It is not difficult to show why there must be someone missing. Multiply the characters together:

2 x 3 x 5 x 7 x 11 x 13

Now here comes the moment for me that is like a twist in this short story that leads to a thrilling and unexpected denouement. What if I were to add 1 to this number?

$$2 \times 3 \times 5 \times 7 \times 11 \times 13 + 1$$

This new number which I've constructed out of the principal characters must in turn be built by multiplying primes together. Remember that was the familiar setting from which we embarked on our journey. So which primes will divide into this new number we've built? Well, it can't be any of the primes in our *dramatis personae*. They will all leave remainder 1. But there must be some primes which divide this number, so this means we must have missed them when we laid out our *dramatis personae*. It turns out this new number is built by multiplying the primes 59 and 509 together.

You might suggest that we add these new characters to our *dramatis personae*, but the beauty of this story is that it can be told again, only to reveal that we will still be missing a character. The revelation is that any finite list of primes will always be missing some characters. Therefore the primes must be infinite.

QED, as mathematicians like to say at the end of their stories.

Tales of the unexpected

What is important to me about a piece of mathematics is not the QED, or final result, but the journey I've been on to get to that point, just as music is not about the final chord. It is certainly important to know that there are infinitely many primes, but our satisfaction comes from understanding why. The joy of reading and creating mathematics comes from that exciting 'aha' moment we experience when all the strands come together to resolve the mystery. It is like the moment of harmonic resolution in a piece of music or the revelation of a murder mystery.

That element of surprise is an important aspect of mathematics. Here is mathematician Michael Atiyah describing the qualities he most enjoys in mathematics: 'I like to be surprised. The argument that follows a standard path, with few new features, is dull and unexciting. I like the unexpected, a new point of view, a link with other areas, a twist in the tail.' When I am creating a new piece of mathematics, the choices I make will be motivated by a desire to take my readers on an interesting journey full of twists and turns and surprises. I want to tease my audience with the challenge of why two seemingly unconnected characters should have anything to do with one another. And then, as the proof unfolds, there is a gradual realisation, or a sudden moment of recognition, that these two ideas are actually one and the same.

One of my favourite theorems is Fermat's discovery of a very curious feature of certain types of prime numbers. If the prime number has the property that when divided by 4 it has a remainder of 1, he believed you could always write that prime number as two square numbers added together. For example, 41 is prime and when I divide by 4 gives you a remainder of 1. And sure enough 41 can be written as $25 + 16$ which is $5^2 + 4^2$. But can this really be true of all such primes? There are infinitely many primes that have remainder 1 when divided by 4. Why should they have anything to do with square numbers?

My initial reaction upon hearing the opening of this story was disbelief. But as Fermat takes me on the journey of his proof I get huge satisfaction as I begin to see these two contrary ideas, primes and squares, being woven together until they fuse into one. It is like a piece of music with two contrasting themes that are varied and developed in such a way that eventually they fuse into one theme.

A simpler example of this idea can be seen with the following little game that I mentioned in Chapter 9. What happens if you add together consecutive odd numbers?

$$1 + 3 = 4, 1 + 3 + 5 = 9, 1 + 3 + 5 + 7 = 16, 1 + 3 + 5 + 7 + 9 = 25$$

Add together N consecutive odd numbers and you will get the Nth square number. Why? The proof is captured by the following picture.

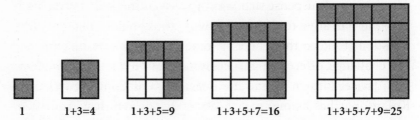

| 1 | 1+3=4 | 1+3+5=9 | 1+3+5+7=16 | 1+3+5+7+9=25 |

The satisfaction comes from the unexpected journey from odd numbers to square numbers. I'm after that 'aha' moment when I suddenly see why there is a connection between these two apparently unrelated characters.

This quality of searching for unexpected connections is one of the reasons why I love talking about one of my own contributions to the mathematical canon: the discovery of a new symmetrical object whose contours have hidden in them potential solutions to elliptic curves, one of the great unsolved mysteries of mathematics. The proof I weave for my fellow mathematicians in my seminars and laid out in my journal article shows how to connect these two disparate areas of the mathematical world.

The joy in telling this story is seeing the moment in my colleagues' faces when they suddenly understand how these two seemingly unconnected ideas could be entwined. The art of the mathematician is not just to churn out the new, but to tell a surprising story. As Poincaré said, it is to make choices.

Just as you sometimes feel a sense of sadness when you come out the other side of a great novel, the closure of a mathematical quest can have its own melancholy. We had been so enjoying the journey Fermat's equations had taken us on that there was a sense of disappointment mixed with the elation that greeted Andrew Wiles's solution to this 350-year-old enigma. That is why proofs that open up the ground for new stories are so highly valued.

The narrative art of mathematics

The quality of suspense that we enjoy in mathematical proofs is a classical narrative tool. Authors will use plot elements that raise questions to keep the reader reading on in the hope of resolving the challenge offered at the beginning of the story. This narrative device, known as the hermeneutic code, was identified by Roland Barthes as one of five key codes present in narrative. It corresponds to unanswered questions or enigmas that demand explication and is an absolutely central device in the creation and execution of a satisfying mathematical proof. What gives us such pleasure when we read mathematics is that desire to have the enigma resolved. In this sense, a mathematical proof shares much in common with a good detective story.

Mathematical proofs all begin with the final scene. The challenge is how we get there. A murder mystery has a similar quality, as does *Star Trek: The Next Generation*'s episode *Cause and Effect*, which opens with the starship *Enterprise* in flames. Picard orders the ship to be abandoned and then we see it exploding. The story begins with the ending. Although most literary narratives don't go for such a dramatic opening, they will be littered throughout with examples of this kind of reverse engineering of cause and effect.

In addition to the tension created by unanswered questions, the other narrative drive in mathematics comes from the action inherent in the proof as it unfolds. In Euclid's proof that there are infinitely many primes, you read that these primes are multiplied together. At once your interest is piqued: OK, so where is this going? What will you do with this new number? The action builds. Oh, you've added one. Even more curious a move. And then the satisfaction of seeing why this sequence climaxes to provide narrative solutions and revelations. This is a good example of the second of Barthes's five codes of narrative, the proairetic code. Suspense is created by the build-up of actions, which by their nature imply further narrative action.

Barthes's other three codes are the semantic, symbolic and cultural. All three dance around the notion that certain concepts in a narrative will resonate with things outside the narrative to give it added meaning. And all three are useful tools to build mathematical proofs, where the reader's pre-existing knowledge is tapped into to ensure the proof has the desired effect. Just as G. H. Hardy spoke of applying a bit of gas, sometimes a proof requires a trigger to ensure that a vast history of ideas will be tapped to progress the proof. Failure to recognise these triggers or references can substantially deplete the effectiveness of the proof, just as it will in the case of a literary narrative.

We often talk about overarching narratives, common to many stories. Some call them masterplots or narrative archetypes. Literary theorists have tried to classify these archetypes and it has been suggested that there are really only seven different types of story. We speak of the Cinderella story or the quest narrative or the battle saga. Does mathematics have its own masterplots? Certainly mathematicians recognise certain proof archetypes and will invoke them to help their readers. There are the proof by contradiction, the probabilistic proof or the proof by induction. The proof of Fermat's Last Theorem depended on creating a world in which the opposite of what you are trying to prove is true. Wiles's proof starts by supposing there were to be a solution to Fermat's equation, and then begins to explore where that would take us. The absurd conclusion to this proposition helps us see that there can't be such a solution.

There is a tension in the best mathematics. Proofs should be neither too complicated nor too simple. The most satisfying proofs have an inevitability about them, yet each step cannot be predicted in advance. In *Adventure, Mystery, and Romance* John Cawelti describes the quality of this tension in literature, but it applies to mathematics too: 'If we seek order and security the result is likely to be boredom and sameness. But rejecting order for the sake of change and novelty brings danger and uncertainty . . . the history of culture can be interpreted as a dynamic tension between the quest for order and the flight from ennui.'

That quest is at the heart of what makes a good proof.

Few professional mathematicians have heard of the Mizar project. Its aim is not one that really interests them. It's building a Library of Babel with everything and nothing inside. And yet I believe machine learning holds a promise that remains untapped. Won't it one day be able to take the mathematics we like and learn to create similar mathematics? Surely this was only a temporary stay of execution.

Music is the creative art most people associate with mathematics. But actually I think it is storytelling that is the closest creative act to proving theorems. I found myself wondering, if mathematical proofs are stories, how good are computers at telling tales?

14

LANGUAGE GAMES

Two scientists walk into a bar.
'I'll have H2O,' says the first.
'I'll have H2O, too,' says the second.
Bartender gives them water because he is able to distinguish
the boundary tones that dictate the grammatical function of
homonyms in coda position, as well as pragmatic context.

Joke posted on Twitter

If you are going to be a writer it's important that you understand language, or at least give the illusion of understanding it. So how good are machines at navigating human communication? The opening sentence of Alan Turing's famous paper 'Computing Machinery and Intelligence' sets out the challenge: 'I propose to consider the question "Can machines think?"' This, Turing believed, was too general, so he refined his challenge: he wondered if a machine could be programmed so that if a human were to engage it in conversation, its responses would be so convincing that the human could not tell it was talking to a machine.

Turing called this the 'Imitation Game', after a parlour game that was popular at the time, but it has become known as the 'Turing Test'. To pass the Turing Test requires an algorithm that can receive as input the vagaries of natural language and process it to produce an output that corresponds to something a human might possibly say in response. ('Natural language' generally refers to language that has evolved naturally in humans through use and repetition without conscious planning or premeditation, in contrast to computer code.)

The first successful effort to take up Turing's challenge was a program called 'ELIZA', developed by the computer scientist Joseph Weizenbaum in 1966. Weizenbaum came up with a clever yet strikingly simple idea. If you've ever been to a therapist trained in person-centred therapy, developed by Carl Rogers, you'll know that the therapist responds with questions that, rather than leading the conversation in a new direction, simply reflect back what the therapist has just been told. This gives clients the space to further elaborate on their state of mind.

> Client: Men are all alike
> Therapist: IN WHAT WAY?
> Client: They're always bugging us about something or
>   other
> Therapist: CAN YOU THINK OF A SPECIFIC EXAMPLE?
> Client: Well, my boyfriend made me come here
> Therapist: YOUR BOYFRIEND MADE YOU COME HERE?
> Client: He says I'm depressed much of the time
> Therapist: DO YOU THINK YOU ARE DEPRESSED MUCH
>   OF THE TIME?

The therapist, in this case, was not a human but a computer program. Weizenbaum understood that it was possible to algorithmically capture the way a Rogerian therapist might respond in a therapy session. He put together a set of 'if . . . , then . . .' rules that took the client's input and, by analysing the structure of the sentence, would output something that depended on the input. For example, if the client says: 'I am X', then ELIZA outputs: 'How long have you been X?' If X = 'sick' then ELIZA's response seems appropriate. But if X = 'sick of you', then the response doesn't quite work. So another rule changes all instances of second-person-singular pronouns to first-person-singular pronouns, and first-person-singular pronouns to second-person-singular pronouns. Now the response becomes: 'How long have you been sick of me?'

The trick is to write enough rules to keep the conversation going convincingly. If the input doesn't match one of the scenarios ELIZA has been programmed to respond to, then it just cleverly invites the client to continue by saying: 'Tell me more.'

Interactions with ELIZA were so convincing that Weizenbaum's own secretary once reportedly asked Weizenbaum to leave the room so that she and ELIZA could have a private conversation. On another occasion a visitor to the lab who had hoped to use the computers for a demonstration found a note on top of the keyboard that read: 'TYPE ALL MESSAGES INTO THIS TELETYPE AND END THEM WITH A PERIOD' (i.e. a full stop). Believing that he was communicating remotely with the person in charge of the lab, he began the following conversation, unaware that he was in fact conversing with ELIZA:

> Visitor: Do you think that I can use the computer this
>   morning?
> Eliza: WHY DO YOU ASK?
> Visitor: I might be able to make some additional Telcom
>   sales.
> Eliza: WHY AREN'T YOU SO SURE?
> Visitor: My perspective [sic] customers have never seen a
>   demonstration of the system.
> Eliza: DOES THIS MEAN VERY MUCH TO YOU?
> Visitor: Of course it does.
> Eliza: YOU SEEM VERY POSITIVE ABOUT THAT.
> Visitor: This is ridiculous.
> Eliza: WHAT DO YOU MEAN, THIS IS RIDICULOUS?

Visitor: Is there any reason why I can't use the computer
now?
Eliza: TELL ME MORE ABOUT YOUR PERSPECTIVE [SIC]
CUSTOMERS.

Exasperated with the evasive answers, he demanded to talk to
the operator in person and typed in angrily:

Visitor: Please dial me up on 491-1850

Because he forgot to end his sentence with a full stop, ELIZA
did not respond. Incensed, he picked up the phone to speak to the
person he thought he'd been conversing with: 'Why are you being
so snotty with me?' he demanded to know. Having just woken up,
the guy in charge of the lab didn't know what the caller was talking about, but when the visitor replayed the conversation back to
him, he couldn't stop laughing. The visitor apparently took some
time to be convinced that he'd been talking to a computer.

That said, it didn't take much conversation with ELIZA to
sniff out that something wasn't quite right about its responses.
They are limited in scope, they are inflexible, and there is no evidence of any memory of the conversation that has taken place.
In a therapy setting it worked quite well because therapy often
depends on the patient having the solution to their problems. An
algorithmic doctor is expected to have the answers, so ELIZA
wouldn't be much good at diagnosing your illness, but there
are algorithms evolving that might be better than your GP at
figuring out what's wrong with you. If you are tempted, you can
give ELIZA a spin with the many online versions that still exist
on the internet.

The challenge of creating a program that could be more flex-
ible and persuasive in its responses was spiced up in the early
1990s by the creation of the Loebner Prize. Underwritten by
Hugh Loebner, an inventor and something of an outsider to the

mainstream research on AI, the prize invited a panel of judges to interact with a human and a computer for a set period of time and decide which was the AI. If a system could fool half the judges, its creator would receive a silver medal and $25,000. In the event that this did not happen in any given year, prizes were to be awarded to the creators of the AI system that convinced most judges it was human.

A gold medal and $100,000 were offered for the first program that judges could not distinguish from a real human in an extended version of the Turing Test that includes deciphering and understanding text, but also would include an audio-visual component. It hasn't been clearly specified what this would entail, but presumably they mean that you get to see and listen to the bot and you've still got to be convinced it's human. Once this is achieved, the annual competition will end. Loebner liked to point out that the gold medal he was offering was solid gold, in contrast to the gold-plated medals of the Olympic Games.

To date no system has won the silver medal, let alone come anywhere near the gold. One of the most recent successes is a chatbot called 'Mitsuku' that goes by the name of Millie and tries to pass itself off as a student from Leeds. It was developed by Steve Worswick, who started programming chatbots as a way to entice visitors to a website that showcased his dance music. It turned out that they much preferred interacting with the chatbot to listening to his music.

In the 2017 competition, which he won, Mitsuku offered some very insightful responses to the judges' questions:

Judge: What do you think of Trump?
Mitsuku: I read that over 50 per cent were either
    concerned or scared about Donald Trump being
    President. If that's the case, why did they vote for him?
Judge: Anything else?
Mitsuku: That is all I have for right now.

However, other responses quickly revealed strange computing glitches:

Judge: What is the answer to 'Add 34957 to 70764'?
Mitsuku: 34957 to 70764 = Ah yes, 34957 is a type of
cheese that is made from sheep's milk. It's not very
popular though.

A strange end to what had been quite a promising conversation up to that point. If you have ever interacted with a chatbot, you have effectively been training the bot with your conversations. Some of these chatbot algorithms work by banking responses that humans make to particular questions, so that if the chatbot encounters a similar scenario in the future it can reverse the roles and sound like the human. But most work on the basis of more sophisticated versions of the ELIZA 'if . . . , then . . .' rules, and these are never really going to be able to handle the varied nature of language. We need something that can grasp how language is put together.

AI systems trying to cope with natural language have difficulty with ambiguity and context. The Loebner Test often concludes with a set of Winograd challenges (named after the Stanford professor who came up with them), which very quickly catch out chatbots that can't untangle inbuilt ambiguity. Take, for example, the word 'they' in the following sentence:

The city councilmen refused the demonstrators a permit
because they [feared/advocated] violence.

The choice of feared or advocated clearly changes depending on what the word 'they' refers to. While a human will know how to unpick this, thanks to context and previous knowledge, machines have an extremely hard time sorting it out. Winograd's sentences exploit the complexity, richness and ambiguity of natural language.

For example, here are some of the Winograd challenges that Mitsuku got in its 2017 Turing Test:

I was trying to open the lock with the key, but someone had filled the keyhole with chewing gum, and I couldn't get it out. What couldn't I get out?

The trophy doesn't fit into the brown suitcase because it's too small. What is too small?

How do we develop the skills to navigate the complexities of language? Our human code is shaped and fashioned by years of verbal interaction with other humans. As children we are exposed to the way language works, we make mistakes, we learn. With the new tools of machine learning, could algorithms finally learn to process natural language? The internet has a huge data set of examples of language in use. So why can't we let an algorithm loose on the internet and learn to navigate the ambiguities inherent in these sentences?

Linguists have been struck by how little language a child needs to hear to be able to understand and interact with other humans. Noam Chomsky sees this as evidence that we are born wired for language. It's as if we were programmed in the old-fashioned top-down model rather than learning from scratch. If that is true, it will be a real challenge for machine learning to pick up language just by being exposed to a huge database of language use.

'This is *Jeopardy!*'

One of the most impressive displays of algorithmic negotiation of the vagaries of natural language came some years ago, a little over ten years after IBM's super computer DeepBlue successfully took the crown from the reigning chess champion Garry

Kasparov. In 2011 IBM turned its attention to a very different sort of competition from chess or Go: it decided to take a shot at the American game show *Jeopardy!*.

Jeopardy! is basically a general-knowledge quiz show. Given that a computer can simply trawl through Wikipedia, that doesn't seem like much of a test for an algorithm. What makes *Jeopardy!* more of a challenge is the style of the questions. They are posed in a sort of inverted manner, where the quizmaster reads something that sounds like the answer to a question and the contestant has to respond with the question. For example, the response to the challenge: 'The name of this element, atomic number 27, can precede "blue" and "green"' is: 'What is "cobalt"'?

Winning at *Jeopardy!* involves understanding a question and accessing a huge database of knowledge to select the most likely answer as quickly as possible. The *Jeopardy!* challenges very often exploit double meanings, plays on words, puns and red herrings, making it very tricky even for humans to unpack the meaning of the question. The ambiguous nature of the questions makes it almost impossible for an algorithm to be 100 per cent accurate. But IBM didn't need to be 100 per cent accurate: it just needed it to be better than other contestants. While some at IBM thought that dedicating time to winning such a trivial game show was a waste of resources, others insisted that success would signal a step change in the ability of machines to parse meaning in language.

If Kasparov was the champion to beat at chess, the *Jeopardy!* kings were Brad Rutter and Ken Jennings, both of whom had notched up extraordinary winning streaks. Jennings had gone seventy-four games in a row unbeaten, while Rutter had earned over four million dollars during his time on the show. Both had cut their teeth on quiz teams at school and university, although Rutter had always been regarded as something of a slacker academically. *Jeopardy!* generally features three competitors, so the two human champions agreed to take on the algorithm that IBM produced. IBM named its algorithm Watson, not after

Sherlock Holmes's sidekick but the first CEO of the company, Thomas J. Watson.

Over two days in January 2011 Rutter and Jennings battled valiantly against Watson and each other. The filming had to be staged at IBM's research lab in Yorktown Heights in New York State because it was impossible to relocate the computer hardware to a TV studio. But, other than the location, everything was set up as normal with the host, Alex Trebek, posing the questions and the shows broadcast on national TV for all to see how close the human race was to being overrun by machines.

The human contestants started off well, and managed to pull ahead at one stage, but in the end couldn't fend off the power of IBM's algorithm. It turned out it wasn't just a matter of being good at answering *Jeopardy!* questions. The quiz show requires a certain amount of game theory, as contestants are given an opportunity to bet money on the final question. This allows a contestant who is behind to bet all their winnings to try to double their money and win the game. Some energy was therefore spent in ensuring that Watson used all its maths skills to wager effectively.

There was one place where Watson seemed to have an unfair advantage: buzzing in. Once a question is read out, contestants have to hit their buzzers to be first to have a chance to answer the question. Originally Watson was going to be allowed to buzz in electronically, rather than having to physically press a button like the humans. But it was soon realised that this would give Watson a huge advantage, so a robotic finger was rigged up that needed to be activated to push the button. Although it slowed Watson down a bit, it was still faster at doing this than humans. As Jennings pointed out: 'If you're trying to win on the show, the buzzer is all.' The problem was, Watson 'can knock out a microsecond-precise buzz every single time with little or no variation. Human reflexes can't compete with computer circuits in this regard.' There was also a certain amount of luck involved:

certain clues on the board score what is called a Daily Double. Watson was fortunate to pick the Daily Double at one point in the game. Had the human contestants lucked out, the game was close enough for Watson perhaps to have lost the match.

Despite Watson's win, it did make some telling mistakes. Having chosen the category 'US Cities', contestants were asked: 'Its largest airport is named for a World War Two hero; its second-largest for a World War Two battle.' The humans responded correctly with 'Where is Chicago?' Watson went for Toronto, a city that isn't even in the United States!

'We failed to deeply understand what was going on there,' said David Ferrucci, an IBM researcher who led the development of Watson. 'The reality is that there's lots of data where the title is US cities and the answers are countries, European cities, people, mayors. Even though it says US cities, we had very little confidence that that's the distinguishing feature.' To its credit, Watson's confidence in the answer was very low (indicated by a whole series of question marks inserted after the response). It was also a question that required wagering money. Watson again indicated a lack of confidence in its answer by placing a low bet.

Responding to the final question, when it was clear Watson had triumphed, Jennings answered: 'Bram Stoker', and then added: 'I for one welcome our new computer overlords'. It was a reference to a popular meme pulled from an episode of *The Simpsons*, itself a spoof of a 1977 B-movie of H. G. Wells's 'Empire of the Ants' (in which a character capitulates in this way to a takeover by giant insects).

Watson showed no sign of understanding the reference.

The way Watson works

The best way to explain how Watson works is to invite you to imagine a huge landscape with words and names and other potential answers scattered about everywhere. The first challenge for IBM was to arrange the words in some coherent manner. The second was to take each question and produce candidate location markers for that question.

Now this is not a three-dimensional landscape such as the one you might see by looking out of the window, but a complex mathematical landscape where different dimensions will be measuring certain properties that will depend on the qualities a word might possess. There is an art to identifying and selecting these qualities. For example, the word might have a strong geographical or chronological association, or be connected to the world of art or sport. It might of course have several of these qualities, in which case its location will be pushed in both of these directions. For example, Albert Einstein would be pushed in the direction of 'scientist' and 'musician', given that he played the violin. But you'd push him more down the scientist dimension than musician. Analysing 20,000 sample questions, the IBM team found about 2500 different answer types, of which some 200 covered over 50 per cent of the questions that were asked.

The Watson algorithm goes through four stages of analysis. First it picks apart the question, to get some idea of where it might lie in the landscape of possible answers. Then it embarks on a process of hypothesis generation, which involves picking some 200 possible answers based on the location of the question. It then scores these different hypotheses. This is done by taking these 200 multidimensional points and crushing them down to points lying along a single line, leading finally to a ranking of possible answers together with some measure of confidence in those answers. If the confidence level passes a certain threshold, the algorithm will buzz in with its proposed answer. All of this has to be done in a matter of seconds, otherwise the human contestants will buzz in first.

Consider a question like:

THE HOLE TRUTH: Asian location where a notoriously horrible event took place on the night of June 20, 1756.

This will score high on the geographical and temporal dimensions. Now, there are presumably a number of Asian locations where something bad happened on 20 June 1756. The word 'hole' in the category will help Watson when it comes to scoring different hypotheses. This means that the Black Hole of Calcutta will be ranked higher than any other Asian location tagged with the same date, giving Watson a winning answer.

The occurrence of words like 'write', 'compose', 'pen' or 'publish' will push you in the direction of artistic creation. So the question 'Originally written by Alexander Pushkin as a poem' would put us in the 'author' region of answers. Once the algorithm selects 200 candidates, the process of scoring these requires a careful weighing of the significance of each of the different dimensions it has picked up. It has to come up with a way of measuring how far the hypothetical answer is from the question. An exact semantic match with a passage in a Wikipedia page might score the answer very highly, but that will have to be combined with other factors. Take the question 'In 1594 he took a job as a tax collector in Andalusia.' The answers 'Thoreau' and 'Cervantes' will both score highly with this semantic match. But then the temporal dimension scores Cervantes higher because his dates, 1547–1616, are closer to 1594 than those of Thoreau, who was born in 1817.

The team working on Watson came up with fifty different scoring components. The algorithm starts with a wide range of candidates because it is unclear at this point what answer will score as highly likely. So the algorithm prefers to include lots of possible answers and to let the scoring pick out the top few. It's a bit like finding a hotel to stay in. First you will select all the hotels within the town or neighbourhood where you want to stay. But then you'll use a scoring system based on price and recommendations that might favour an outlying hotel worth visiting.

The way the algorithm does the scoring allows it to learn from its mistakes in a bottom-up fashion and refine its parameters, a bit like twiddling a dial to redefine the function. The art is trying to find the best setting for your dials to get the right answer in as many different contexts as possible. Consider the question 'Chile shares its longest land border with this country.' Two countries share a border with Chile: Argentina and Bolivia. So how would you score these two hypothetical answers? One might decide to score an option higher if it were to be mentioned more often. In this case, Bolivia would have received a higher score because Chile and Bolivia have had many disputes over their borders that have spilled over into the news. But if you were to score source material of a more geographical nature higher and to count the mentions of each country in these publications, then Argentina would come out on top, which is in fact the correct answer.

When Jennings was told how Watson worked he was quite startled. 'The computer's techniques for unravelling *Jeopardy!* clues sounded just like mine,' he said. Jennings would home in on keywords in a clue and then rake through his memory (Watson had access to a 15-terabyte databank of human knowledge) for clusters of associations with those words. He then carefully reconsidered the top contenders in light of all of the contextual information he could muster: the time, place, gender hinted at in the clue; whether it is sports, literature or politics. 'This is all an instant, intuitive process for a human *Jeopardy!* player, but I felt convinced that under the hood my brain was doing more or less the same thing.'

Why did IBM go to all this effort? Winning games may sound rather pointless but for companies like IBM and DeepMind they offer a very clear indication of success. You either win or lose. There is no room for ambiguity. They provide great publicity stunts for a company that needs to sell its product because every-one loves the drama of human versus machine. They are like an algorithmic catwalk allowing a company to show off its coding prowess.

IBM Watson has already changed our perception of what computers may do – it beat the best *Jeopardy!* champions, and it is being used for medical diagnoses. What sets Watson apart? What makes it different? This capability to take into unstructured data is a big strength for Watson. We train it. Additionally just dumping the text in Watson, humans actually form the system to understand what is most important and reliable inside the text. Watson pulled in all of Wikipedia prior to its *Jeopardy!* appearance, and stored that data offline. Humans can tell Watson to trust one source of info more than another. This shift from scheduling to training is part of why IBM calls this effort cognitive computing.

At the future, we'll rely less on rote calculation, and more on interaction and learning. It is clever enough to know that with a little more info, it'd be capable to rule out an answer, or increase confidence in one of the answers it is already offering. When Watson handles a difficult question in its current applications, it comes back with a set of possible outcomes – but it is also able to ask clarifying questions. Most question-answering systems are programmed to deal with a defined set of question types – meaning you can only answer certain kinds of questions, phrased in a certain ways, in order to obtain a response. Watson handles open-domain questions, meaning anything you can think of to ask it. It uses natural-language processing techniques to pick apart the words you give it, in order to understand the real question being asked, even when you ask it in an unusual way.

IBM actually published a very useful FAQ about Watson and IBM's DeepQA Project, a foundational technology utilised by Watson in generating hypotheses. The computer on *Star Trek* is a more suitable comparison. The fictional computer system can be seen as an interactive dialogue agent that could answer questions and provide precise info on any subject.

Lost in translation

I struggled with learning languages at school and still remember reading in *The Hitchhiker's Guide to the Galaxy* about the Babel fish, a small, yellow, leech-like creature that, when dropped into your ear, would feed on brain waves to instantly translate anything that was said to you in any form of language. Now that sounded really useful! As so often happens, yesterday's science fiction has become today's science fact. Google recently announced the creation of an earpiece called Pixel Buds that does exactly what Douglas Adams dreamed about.

Given that the input is already a well-formed sentence, you might think that the work of navigating language has already been done and a word-for-word exchange will do. But simple word substitutions will often result in an amazing word soup. For example, take this quote from *Madame Bovary*: 'La parole humaine est comme un chaudron fêlé où nous battons des mélodies à faire danser les ours, quand on voudrait attendrir les étoiles.' Taking my French–English dictionary and translating one word at a time (and having to make some choices as there are different possible translations for each word) gives me: 'The speech human is like a cauldron cracked where we fight of the melodies to make to dance the bears, when one would like to tenderise the stars.' Not, I think, what Flaubert had in mind! This is where a sensitivity to the workings of a given language is essential. Once we see that the word 'battons' comes close to the word 'mélodies', we might go for an alternative translation of 'battons' not as 'fight' but as 'beat', and might even add in 'the rhythm'. But that still leaves us with the puzzle of what it might mean to 'tenderise' the stars.

An effective translation algorithm needs to have a good sense of what words are likely to come together. I remember having great fun with my best friend at university, who was studying Persian. When I looked through his Persian–English dictionary, it seemed like every word had at least three completely different meanings, one of which was sexual. We whiled away a lot of time cooking up crazy translations from a single Persian sentence.

Modern translation algorithms are tapping into the underlying mathematical shape of a language. It turns out we can plot words in a language as points in a high-dimensional geometric space and then draw lines between words which have structural relationships to one another. For example, 'man' is to 'king' as 'woman' is to 'queen' translates mathematically into the fact that if you draw lines between these pairs of words they will be parallel and will point in the same direction. You end up with a shape that looks like a high-dimensional crystal. The interesting thing is that French and English have very similarly shaped crystals, so you just have to figure out how to align them.

I put Flaubert's line from *Madame Bovary* into Google Translate to see how well it would capture its meaning. We get pretty close with: 'The human word is like a cracked cauldron where we beat melodies to make the bears dance, when we want to soften the stars.' The word 'soften' is certainly better than 'tenderise' but it still doesn't quite ring true. Turning to my OUP translation (still done by humans, in this case Margaret Mauldon), I find: 'Human speech is like a cracked kettle on which we tap crude rhythms for bears to dance to, while we long to make music that will melt the stars.'

You realise how important it is not only to choose the right word but to capture the sentiment of the sentence. The algorithmic translators are still tapping out crude rhythms for bears to dance to while humans can translate prose that gets closer to melting the stars. Most of the time tapping out crude rhythms will be good enough provided that the meaning, if not the poetry, of the sentence is communicated. As evidence of its success, Google Translate currently supports 103 languages and translates over 140 billion words every day.

But how soon will it be before human translators and interpreters are put out of a job or at least reduced to fixing glitches in computer translation rather than producing fresh text? My feeling is that actually these algorithms will never reach the status of human translation. At least not until AI has cracked the problem of consciousness. Translation is more than just moving from one language to another. We are moving from one mind to another, and until there is a ghost in the machine it will never be able to fully tap into the subtlety of human communication.

Looking back over both translations of the *Madame Bovary* text, I actually quite like Google's suggestion of 'cauldron' rather than 'kettle'. And 'to make the bears dance' has a slightly more menacing feel than the human translation. Perhaps it's the combination of human and machine that ultimately might give the best translation.

That is why, to get more nuanced translations, Google has enlisted human helpers to improve its algorithm, but this doesn't always lead to better outcomes. Some people can't resist messing with the algorithm, as was illustrated when Google started translating Korean headlines about Kim Jong-un, the leader of North Korea, by referring to him as Mr Squidward, a character from *SpongeBob SquarePants*. Hackers had managed to suggest enough times that Mr Squidward was a better alternative translation for 'supreme leader', the term used by the North Korean media to refer to Kim Jong-un, to trip the algorithm. They changed the probabilities by loading the data with false examples. A similar hack occurred when the official title for the Russian Federation was translated into Ukrainian as Mordor (the land occupied by *The Lord of the Rings*' evil Sauron).

Despite these glitches, Google Translate is getting ever more adept at moving from one human language to another. There is even a proposal to map the sound files of animal communications and see if their multidimensional crystal might have the same or a similar shape to human communication, allowing us to understand what our pets are saying. Soon we may need a new tool to help us understand the languages emerging from machines, or so I began to think after witnessing an amazing act of linguistic creativity at the Sony Computing Laboratory in Paris, where Luc Steels has created robots that can evolve their very own language.

Robot lingo

Steels suggested that I come to visit his lab, where twenty robots had been placed one by one in front of a mirror and invited to explore the shapes they could make using their bodies in the mirror. Each time they came up with a new shape, they created a new word to describe it. For example, the robot might choose to name the action of putting its left arm in a horizontal position. Each member of the population of robots created its own unique language for its own unique set of actions.

The really exciting part came when these robots began to interact with one another. One robot chose a word from its lexicon and asked the other robot to perform the action corresponding to that word. Of course, the second robot hasn't a clue what it is asking for. So it chooses one of its positions as a guess. If it has guessed correctly, the first robot confirms this. If not, it shows the second robot the intended position.

The second robot may have given this action its own name, in which case it won't abandon its choice but will update its dictionary to include this new word. As its interactions progress, the robot weights the value of the words according to how successful the communication has been, downgrading those where the interaction failed. The extraordinary thing is that after a week together a common language began to emerge. By continually updating and learning, the robots developed their own language. It was sophisticated enough to include words that represent the more abstract concepts of 'left' and 'right'. These words evolved on top of the direct correspondence between word and body position. The fact that there should be any convergence at all is exciting, but the really striking fact for me was that these robots had a new language which they understood yet the researchers at the end of the week could not comprehend until they had interacted with the robots enough to decode the meaning of these new words.

Steels' experiment offered a beautiful proof of how Ada Lovelace was wrong. Steels had written the code that allowed the robots to generate their own language, but something new had emerged from the code, demonstrated by the fact that no one other than the robots could understand their common language. The only way to learn this language was for the robot to demonstrate what body position corresponds to each sound.

Google Brain has pushed this idea of algorithms creating their own language to develop new methods of encryption, so that two computers can talk to one another without a third being able to eavesdrop. In the cryptographic world Alice is tasked with sending Bob secret messages that Eve will try to crack. Alice scores points if Eve can't decrypt her message, and Eve scores if she can. Alice and Bob start by sharing a number, which is the only thing Eve doesn't have access to. This number is the key to the code they will create. Their task is to use this number to create a secret language that can only be decrypted if you know the key.

Initially Alice's attempts to mask the messages were easily hacked. But after 15,000 exchanges Bob was able to decrypt the messages Alice sent, whereas Eve was scoring at a rate that was no better than if she had been randomly guessing the message. It wasn't just Eve that was shut out. The neural networks Alice and Bob were using meant that their decisions were very quickly obscured by the constant reparameterising of the language, so that even by looking at the resulting code it was impossible for humans to unpick what they were doing. The machines could speak to one another securely without us humans being able to eavesdrop on their private conversations.

Stuck in the Chinese Room

These algorithms that are navigating language, translating from English to Spanish, answering *Jeopardy!* questions and comprehending narrative raise an interesting question which is important for the whole sphere of AI. At what point should we consider that the algorithm understands what it is actually doing? This challenge was captured in a thought experiment created by John Searle and called 'The Chinese Room'.

Imagine that you are put in a room with an instruction manual which gives you an appropriate response to any written string of Chinese characters posted into the room. With a sufficiently comprehensive manual, I could potentially have a very convincing discussion with a Mandarin speaker without ever understanding a word.

Searle wanted to demonstrate that a computer programmed to respond with text that we would struggle to distinguish from a human respondent cannot be said to have intelligence or understanding. Embedded in this line of thought is a powerful challenge to Turing's test. But, then again, what is my mind doing when I'm articulating words now? Am I not at some level following a set of instructions? Might there be a threshold beyond which we would have to regard the computer as understanding Mandarin?

And yet when I refer to a chair I know what I am talking about. When a computer has a conversation about a chair it has no need to know that this thing 'chair' is a physical object which people sit on. It just has to follow the rules, but following rules does not constitute understanding. It will be impossible for the algorithm to achieve perfect use of the word 'chair' if it hasn't experienced a chair. And this is why the question of embodied intelligence is one that is particularly relevant to current trends in AI.

In some ways language is a low-dimensional projection of the environment around us. As Franz Kafka said: 'All language is but a poor translation.' All physical chairs are different, yet they are compressed into one data point in language: chair. But this data point can be unwrapped by another human into all the chairs it has experienced. We can speak of an armchair, a bench, a wooden chair or desk chair and these all bring up different specific associations. These are the word games Wittgenstein famously talked about. A computer without embodiment is stuck in the low-dimensional space of Searle's room.

It comes down to the strange nature of consciousness, which allows us to integrate all of this information into a single unified experience. If we take an individual neuron, it doesn't understand English, and yet at some point, as we build up the brain neuron by neuron, we see that it does understand language. When I am sitting in the room processing the incoming Mandarin with my manual, I am acting like part of the brain, a subset of neurons responsible for the processing. Although I don't understand what I'm saying, maybe we should say that the entire system made up of the room containing me and the manual does understand. It's the complete package that makes up the whole brain, not just me sitting there. In Searle's room I'm more like the computer's CPU, the electronic circuitry that carries out the instructions of a computer program by performing the basic calculations.

Could a computer form sentences of meaning – or even beauty – without understanding language or being exposed to the physical world around it? This is a question programmers are grappling with right now in a variety of ways. Maybe a machine doesn't need to understand what it's saying in order to produce convincing literature. Which brings me back to the question that set me off on this excursion into language in the first place: how good is modern AI at taking language and weaving the words together to tell a story?

15

LET AI TELL YOU
A STORY

A man who wants the truth becomes a scientist;
a man who wants to give free play to his subjectivity
may become a writer; but what should a man do
who wants something in between?

Robert Musil

Some of the stories I grew up on have left a lasting impression. High on the list is Roald Dahl's *Tales of the Unexpected*, with its unnerving account of a man who eats so much royal jelly he turns into a bee, a tramp tattooed by a famous artist who sells his skin to the highest bidder, an obedient housewife who kills her husband with a frozen leg of lamb and serves it to the detectives investigating the case. One of these disturbing tales, written in 1953, tells the story of 'The Great Automatic Grammatizator'.

The mechanically minded Adolphe Knipe had always wanted to be a writer. Alas, his efforts were hackneyed and uninspiring. But then he had a revelation: language follows the rules of grammar and is basically mathematical in principle. With this insight he set about creating a mammoth machine, the Great Automatic Grammatizator, able to write prize-winning novels based on the works of living authors in fifteen minutes. Knipe blackmails these authors into licensing their names rather than having it revealed that writing a novel is something a machine can do easily and often better. As the story ends, the narrator is wrestling with his conscience:

> This very moment, as I sit here listening to the crying
> of my nine starving children in the other room, I can feel
> my own hand creeping closer and closer to that golden
> contract that lies over on the other side of the desk.
> Give us strength, Oh Lord, to let our children starve.

Roald Dahl died before such a machine was within the realm of possibility, but suddenly it no longer seems such a crazy idea.

One of the very first programs written for a computer was developed to write love letters. After cracking the Enigma code at Bletchley Park, Alan Turing headed to the University of Manchester to put into practice his ideas for a physical version of the all-purpose computer he'd been theorising about. Under his guidance the Royal Society Computing Laboratory soon produced the world's first commercially available general-purpose electronic computer, the Ferranti Mark 1. It was used to find new primes, wrestle with problems in atomic theory and explore early genetic programming.

Members of the team were perplexed when they began to find letters of the following ilk lying around the lab:

> DUCK DUCK
>   you are my wistful enchantment. my passion curiously
> longs for your sympathetic longing. my sympathy
> passionately is wedded to your eager ambition. my
> precious charm avidly hungers for your covetous ardour.
> you are my eager devotion.
> yours keenly
>   M. U. C.

MUC was the abbreviation for Manchester University Computer. Christopher Strachey, an old friend of Turing's from his days at King's College, Cambridge, had decided to see if the Ferranti Mark 1 might be able to tap into a more romantic side of its character. He had taken a very basic template:

YOU ARE MY [adjective] [noun]. MY [adjective] [noun] [adverb] [verbs] YOUR [adjective] [noun].

Strachey programmed the computer to select words at random from a data set he had cooked up and insert them into the variables in his simple algorithm. The randomness was achieved using a random-number generator that Turing had built for the computer. Anyone receiving more than one or two of these mystifying love letters would soon spot a pattern and deduce that their Valentine was unlikely to sweep them off their feet.

Algorithmically generated literature is not new. A whole school of writers and mathematicians came together in France in the 1960s to use algorithms to generate new writing. The group called itself Oulipo, for *Ouvroir de littérature potentielle*, which roughly translates as 'workshop for potential literature'. Raymond Queneau, one of the founders, believed that constraints were an important part of the creative process. 'Inspiration which consists in blind obedience to every impulse is in reality a sort of slavery,' he wrote. By imposing quasi-mathematical constraints on writing, he felt you could achieve a new sort of freedom. The group's early projects focused on poetry. As anyone who has written a poem knows, the constraints of poetry will often push you into new ways of expressing ideas that free-form prose would never have unearthed.

One of the group's most popular algorithms, conceived by Jean Lescure, is S + 7 (or, in English, N + 7). The algorithm takes as its input any poem and then acts on all the nouns in the poem by shifting them seven words along in the dictionary. The S stands for *substantifs*, which is French for 'nouns'. The output is the ensuing rewritten version of the original poem. For example, Blake's poem:

> To see a World in a Grain of Sand
> And a Heaven in a Wild Flower
> Hold Infinity in the palm of your hand
> And Eternity in an hour.

becomes:

> To see a Worm in a Grampus of Sandblast
> And a Hebe in a Wild Flu
> Hold Inflow in the palsy of your hangar
> And Ethos in an housefly.

Lescure hoped this curious exercise would prompt us to revisit the original text with new eyes and ears. The algorithm changes the nouns but keeps the underlying structure of the sentences, so it perhaps could help reveal structural elements of language masked by the specific meaning of the words.

Queneau, who had studied philosophy and was a member of the Mathematical Society of France, was fascinated by the links between mathematics and creativity. He sought to experiment with different ways to generate new poetry using the tools of maths. Shortly before founding Oulipo he had composed a book of sonnets which he called *100,000,000,000,000 Poems*. Ten different versions were proposed for each line. There were thus ten choices for the opening line and ten choices for the second line, making a total of 100 different possibilities for the first two lines. Given that there are fourteen lines in a sonnet, that makes a possible 10^{14} different poems in total. That's a hundred thousand billion new sonnets! If the first diplodocus ever to have evolved during the Jurassic period had started reciting Queneau's sonnets at one a minute, it would have got through all the possibilities only once by now.

Queneau had cooked up a literary version of Mozart's game of dice. Chances are the following sonnet, which I picked at random, has never appeared in print before:

Don Pedro from his shirt has washed the fleas
His nasal ecstasy beats best Cologne
His toga rumpled high above his knees
While sharks to let's say potted shrimps are prone
Old Galileo's Pisan offerings
Nought can the mouse's timid nibbling stave
He's gone to London how the echo rings
The nicest kids for stickiest toffees crave
Emboggled minds may puff and blow and guess
In Indian summers Englishmen drink grog
And played their mountain croquet jungle chess
We'll suffocate before the epilogue
Poor reader smile before your lips go numb
Fried grilled black pudding's still the world's best yum

As the Oulipo movement illustrates, poetry is particularly amenable to an algorithmic approach. The constrained nature of the form provides a template that the algorithm can try to fill in a meaningful manner. A pattern is chosen, a haiku or a sonnet, and the task of the algorithm is to choose words to match the pattern while attempting to come up with some form of overarching coherence. Whenever I've attempted to write poetry with a rhyming pattern, I've found it useful to tap into a database of words that rhyme. Weaving a line through the constraints of rhyme and rhythm is something a computer can do in spades.

That is the principle behind the code underpinning the Cybernetic Poet, a more recent creation of the futurist Ray Kurzweil, who writes frequently on the impending fusion between man and machine. Rather than relying on words randomly picked out of a dictionary, Kurzweil trained his Cybernetic Poet on the work of accomplished poets like Shelley and T. S. Eliot. Here is one of the Cybernetic Poet's haikus, informed by a reading of Keats:

You broke my soul
The juice of eternity,
The spirit of my lips.

Although the poem does indeed have seventeen syllables, the algorithm seems to have missed that a haiku should divide symmetrically into five syllables for the first line, seven for the next and five for the final line.

Here is a poem that fuses Shelley and Eliot:

Lady of Autumn's being,
Thou, from the day, having to care
Teach us now thoroughly small and create,
And then presume?
And this, and me,
And place of the unspoken word, the unread vision in
    Baiae's bay,
And the posterity of Michelangelo.

'Ode to the West Wind' meets 'The Love Song of J. Alfred Prufrock'.

In a Turing Test conducted by Kurzweil, the Cybernetic Poet was able to trick human judges most of the time. This is partly because gnomic outputs are part of the landscape of modern poetry, leaving the reader to do much of the work of interpretation. An enigmatic output from an algorithm can pass for poetry written by a human. The results and poems Kurzweil used can be found on his website: http://www.kurzweilcyberart.com/.

If you'd like to have a go at distinguishing human poetry from the efforts generated by a range of algorithms, Benjamin Laird and Oscar Schwartz have put together a challenging poetic Turing Test in a project they've called 'bot or not' at http://botpoet.com.

The Cybernetic Poet might be doing well at producing convincing poetry, but creating a Cybernetic novelist is a much taller order.

How to write a novel in a month

Lescure's idea of applying algorithms to existing literature is a trick that has been exploited by a number of coders who have taken part in NaNoGenMo (National Novel Generation Month), a response to the National Novel Writing Month, which invited budding authors to knock out 50,000 words in the month of November. The software developer and artist Darius Kazemi decided that instead of going to the trouble of cranking out 1667 words a day, he would spend the month writing code that could generate a 50,000-word novel. His plan was to share both the novel and the code at the end. His tweet about his idea in 2013 started an annual literary hackathon.

Many of the coders who have taken part in NaNoGenMo have relied on perturbing existing texts: *Pride and Prejudice* run through a Twitter filter; *Moby Dick* interpreted through a sci-fi algorithm; Gustavus Hindman Miller's *Ten Thousand Dreams, Interpreted* reinterpreted and reordered by code. But it is a more ambitious work called *The Seeker* that caught people's attention. The novel documents an algorithm's struggle to understand how humans operate by reading different articles on wikiHow. The algorithm has a meta-code of Work, Scan, Imagine, repeat. The author of the code, who goes by the name of 'thricedotted', tells us what this means:

In *Work* mode, it scrapes concepts about human activities.
In *Scan* mode, it searches plain text 'memories' from a

seed concept encountered during *Work*. It then uses the concepts it didn't recognize from *Scan* mode (censored out in its logs) to *Imagine* an 'unvision' around the seed concept.

The Seeker chronicles the algorithm's journey of discovery as it explores the database of wikiHow, building from ignorance to some semblance of understanding. The first 'how to' page it consults is 'How to get a girl to ask you out'. The seed picks up from this scan the word 'hurt', which is mentioned in relation to how not to hurt a girl's feelings. In its Imagine mode it then produces a surreal riff on the word 'hurt'.

The Seeker almost works, unlike many other algorithmic novels, because you start to feel you are getting inside the head of the algorithm as it tries to make sense of humans. The fact that the output reads like a strange computer code of words is consistent with the algorithm's potential internal voice. This may in fact be the ultimate goal of any algorithmically generated literature: to allow us to understand an emerging consciousness (if it ever does emerge) and how it differs from our own.

But for now the commercial world would be content with algorithms that could knock out the next Mills and Boon romance or Dan Brown thriller. Many of these bestsellers are based on clear-cut formulae. Couldn't someone simply automate that formula? If algorithms can't produce great works of literature, maybe they could churn out commercial staples like Ken Follett or even an algorithmic *Fifty Shades of Grey*. An algorithm written by a commissioning editor, Jodie Archer, and a data analyst, Matthew Jockers, does at least claim to spot whether a book is likely to be a bestseller. The algorithm found that readers of bestsellers liked shorter sentences, voice-driven narratives and less erudite vocabulary than readers of literary fiction. If only I'd known that before I started!

Harry Potter and the deathly Botnik

Most of the examples I've pointed to so far rely on a top-down model of programming: a poetry template filled in randomly following an explicit set of rules; code that transforms classic texts into new work; algorithms that are programmed to take data and turn them into stories. These programs don't really allow for much freedom. Machine learning is changing that. It's now possible for an algorithm to take an author's entire opus and learn something about the way they write. If they favour a particular word, there may be a high probability that this word will be followed by certain other words. By building up a probabilistic picture of how an author uses words, an algorithm could start to generate the continuation of a text. This is how predictive texting works. The literary results have been both revealing and entertaining.

This use of machine learning to create new literature has been championed by a group that calls itself Botnik. Founded in 2016 by the writer Jamie Brew and the former cartoon editor of the *New Yorker*, Bob Mankoff, Botnik is now an open community of writers who use technology in the creation of comedy. The group has taken *Seinfeld* scripts and produced new episodes based on a mathematical analysis of past dialogue and even got an actor from *Scrubs*, Zach Braff, to perform a monologue produced by Botnik based on the medical comedy drama. The result is sometimes surreal. In the Botnik *Seinfeld* episode, Jerry confidently declares: 'Dating is the opposite of tuna, salmon is the opposite of everything else. I'm sure you know what I mean.'

Botnik has also taken Thanksgiving recipes and produced a YouTube video to take people through the dinner you'd get if you left the cooking to an algorithm:

The best way to make something really special for
Thanksgiving is to fold the turkey in half and then just
throw it right in the kitchen.

Probably their most successful output to date came from training Botnik on the seven volumes of *Harry Potter*. The three pages
it generated have a very convincing ring to them.

Magic: it was something that Harry Potter thought was
very good. Leathery sheets of rain lashed at Harry's ghost
as he walked across the grounds towards the castle.

But there are moments of pure genius that could only have
come from an algorithm:

Ron was standing there and doing a kind of frenzied
tap dance. He saw Harry and immediately began to eat
Hermione's family. Ron's Ron shirt was just as bad as
Ron himself.

I guess for fans who are really desperate for more from the
wizarding world this may be better than nothing, but it's pretty
plot-free and is unlikely to sustain much drama beyond three
pages.

I decided to investigate whether if I fed Botnik the data of my
first book, *The Music of the Primes*, it would provide me with a
new insight I might have missed. I got the following strange take
in response:

The primes are the jewels which shine amongst the
vast expanse of our infinite universe of numbers. As he
counted higher and higher Gauss suddenly saw a pattern
beginning to emerge. His passion for the problem was
further fuelled when his father offered to buy him a
Ferrari. Previously education schemes had been geared
to the creation of each list of primes 2, 3, 5, 7, 11 and
13 years respectively. For all but their last year they
remain in the ground feeding on the sap of tree roots.

A bizarre but recognisable mash-up of my first book. One of the important lessons I learned from applying this algorithm is that there is still significant human involvement in creating texts. What the algorithm does is give you a choice of eighteen words that are likely to follow given what exists to date. But that gave me a lot of freedom to take it in whatever direction tickled my fancy. Often the human component of artistic creations by algorithms is masked. It makes for a better story to say: 'AI writes new *Harry Potter*!' than 'Another writing student has produced a new novel'.

I think it's fair to say that novelists are not likely to be put out of a job any time soon. Maybe all that Botnik is capturing is the fact that authors do have a style, which is recognisable from the way they construct their sentences. But it is only capturing that: the local evolution of text. There is no attempt to reproduce a global narrative structure. It is like the jazz Continuator: it can produce a few phrases of convincing jazz but ultimately becomes boring as it doesn't know where it is going. I often wonder whether algorithms are already at work at Netflix and Amazon, knocking out scripts that keep us watching but ultimately take us nowhere.

What if . . . ?

The storytelling algorithm Scheherazade-IF, developed by Mark Riedl and his colleagues at the Georgia Institute of Technology, was set up in 2012 to tackle this deficit. Its goal is to navigate a more coherent pathway through the maze of possible stories. The algorithm owes its name to the famous storyteller Scheherazade, who saved her life by coming up with new stories, night after night, to enthral and distract her murderous husband. (The IF stands for 'Interactive Fiction'.) If you ask Scheherazade-IF to construct a story about a specific subject or situation it hasn't encountered before, it will learn about it by sourcing and digesting previous stories.

'Humans are pretty good storytellers and possess a lot of real-world knowledge,' says Riedl, one of the lead developers of the algorithm. 'Scheherazade-IF treats a crowd of people as a massively distributed knowledge base from which to digest new information.' It then compiles these examples into a tree of possible directions in which the story could go based on these previous stories. This kind of skill is really useful when it comes to more open-ended computer games, where you might want many different possible scenarios within the game play. A good storyteller will find the best route through the tree of possible stories.

This taps into a genre of storytelling I used to love as a kid. In Gamebooks or Choose Your Own Adventure you are given choices at certain points in the narrative: turn to page 35 if you want to go through the left door, page 39 if you want to go through the right door. The trouble is, your choices will sometimes produce rather incoherent stories. Given that a story with just ten junctions could produce over 1000 different stories, you'd like some way for an algorithm to find the best ones.

Scheherazade-IF is trying to do exactly this with the tree of possible scenarios it has generated from its data-gathering on the web. So how good is it at choosing a satisfying path? Tests by the research team suggested it chose pathways that were rated as being as good as human-chosen pathways and it scored much higher than a randomly generated journey. The algorithm was able to make far fewer logically inconsistent moves than the randomly generated tale. Logical inconsistency is something that immediately gives away the fact that a piece of writing is generated by an algorithm. You don't want to find that a character killed off in Chapter 2 suddenly reappears in Chapter 5 (unless it's a zombie story, I guess).

It's all well and good to trawl the web for old stories and put them together afresh, but what about the challenge of imagining scenarios that have never been cooked up before? This was the goal of the What If Machine, or Whim, funded by the EU. One of the problems authors face when trying to create something new is that they will get stuck in bounded ways of thinking. The What If Machine tries to take storytellers out of their comfort zones by suggesting new possible scenarios.

This is, of course, what we do all the time when we want to create a new story: 'What if a horse could fly?' and you've got Pegasus. 'What if a portrait of a young man aged while he himself stayed young?' and you've got *The Picture of Dorian Gray*. 'What if a girl suddenly found herself in a strange land where animals could talk and everyone was mad?' and you've got *Alice's Adventures in Wonderland*. Many of Roald Dahl's *Tales of the Unexpected*, which I so loved as a kid, exploit the 'What if . . . ?' model of creativity.

In fact storytelling in humans probably has its genesis in the question 'What if . . . ?' Storytelling was our way of doing safe experiments. By telling a 'What if . . . ?' story, we are exploring possible implications of our actions. The first stories probably grew out of our desire to find some sort of order in the chaos surrounding us, to find meaning in a universe that could be cruel and senseless. It was an early form of science. Sitting around the fire sharing stories of the day's hunt helped the tribe be more successful the next day. What *Homo sapiens* lacked in strength they made up for in the collective strength of the tribe. That strength grew with increased socialising and sharing. The fire of the campsite appears to be what lit the spark of creativity in humans.

Whim was hoping to ignite creativity around the digital fireside. One of Whim's first ventures took as its starting point the idea of Pegasus: a horse that could fly. Could an algorithm come up with other curious animals that might stimulate a story? It started with a database of animals and listed all of the properties they might have. The National Geographic Children's website was a good place to start. The website tells you that a dolphin is a mammal that lives in the sea and can be ridden by humans. A parrot is a bird that can fly and sing. But if you now get your algorithm to start mixing and matching, you might get a flying mammal that humans can ride and that sings: something that could easily appear in a fairy tale or a volume of *Harry Potter*.

The principle is similar to those books where you have a head, a torso and legs that you can mix up to create strange combinations. With ten choices for each body part, that would give you 1000 different animals. But if your list is going to be useful, you will have to come up with some way to evaluate the ideas. The team at Whim introduced mathematical functions that score the suggestions for stimulation and novelty and seek to reject any idea that is too vague to be helpful. This led to some interesting suggestions bubbling to the top:

An animal that has eyes with which it can defend itself
A tiger with wings
A bird that lives in a forest that can swim under water

New animals with strange skills are a good catalyst for new stories. The next step was to program Whim to generate novel narrative ideas. It started by taking a series of 'What if . . . ?' storylines that we would immediately recognise and then perturbed the assumptions implicit in these scenarios. The hope was that this would spark creativity by combining topics in surprising and subversive ways. Whim is programmed to generate narrative suggestions in six fictional categories: Kafkaesque, Alternative Scenarios, Utopian and Dystopian, Metaphors, Musicals and Disney. The results are varied in their success.

In the Disney section, Whim came up with a storyline that could conceivably find itself in the next *Inside Out*: 'What if there were a little atom who lost his neutral charge?' That may be one for the geeks among us. Some of the Disney suggestions might be better categorised as a little dystopian: 'What if there were a little plane that couldn't find the airport?'

The following storyline, from the Alternative Scenarios category, was distinctly less promising: 'What if there were an old refrigerator who couldn't find a house that was solid? But instead, she found a special style of statue that was so aqueous that the old refrigerator didn't want the solid house any more.' Or this from Kafkaesque: 'What if a bicycle appeared in a dog pound, and suddenly became a dog that could drive an automobile?'

The What If Machine was responsible for suggesting a storyline that eventually led to the staging of a West End musical in 2016. The TV channel Sky Arts, interested in probing the limits of algorithmic creativity, had commissioned a musical created by AI. It filmed the process of development and eventually staged it. In order to come up with a scenario for the musical, Whim was brought on board. The algorithm came up with a range of different scenarios, which were then passed through another algorithm, developed in Cambridge. This second algorithm had analysed the storylines of musicals to learn what makes a hit and what flops and was tasked with choosing one of Whim's suggestions for further development. It picked out the following as a potential hit: *'What if there were a wounded soldier who had to learn how to understand a child in order to find true love?'*

At this point another algorithm, PropperWryter, which had had some success in generating fairy tales, took over. Its fairy-tale algorithm was trained on thirty-one narrative archetypes for Russian folktales identified by the structuralist Vladimir Propp in 1928. PropperWryter developed the plot of the scenario provided by Whim and turned it into a story about the Greenham Common women's anti-nuclear movement. The music was provided by yet another algorithm called Android Lloyd Webber.

Beyond the Fence hit the West End for a short run at the Arts Theatre in the spring of 2016. To realise the project there was probably as much human intervention as computer creativity.

The result was not much of a threat to Andrew Lloyd Webber. As the theatre critic Lyn Gardner summed up in her two-star review: 'a dated middle-of-the-road show full of pleasant middle-of-the-road songs, along with a risibly stereotypical scenario and characters'. But then maybe what we should really take away from this is that reviewers aren't particularly well disposed to giving algorithms too much credit.

The Great Automatic Mathematizator

Asking 'what if . . . ?' is not far from the way a mathematician pushes the boundaries of knowledge. What if I imagine there is a number whose square is −1? What if I imagine there might be geometries where parallel lines meet? What if I twist a space before I join it up? The idea of perturbing known structures to see if anything worthwhile emerges from the variations is a classic tool in developing new mathematical narratives. Could a mathematical 'What if . . . ?' algorithm actually help in making new maths? If mathematics is a kind of storytelling with numbers, how effective are current algorithms at generating new mathematical tales?

Simon Colton, who wrote the code behind the Painting Fool and is the co-ordinator of Whim, joined forces with Stephen Muggleton at Imperial College London to explore exactly this question. They developed an algorithm that would take accepted mathematics and see if they could prompt new ideas. Colton let the algorithm loose on one of the most visited mathematical websites on the internet, The On-Line Encyclopaedia of Integer Sequences, a project initiated by Neil Sloane to collect all the interesting sequences of numbers and figure out how they are generated. It includes old favourites like:

1, 1, 2, 3, 5, 8, 13, 21 . . .

which anyone who has read *The Da Vinci Code* will recognise as the famous Fibonacci numbers. They are generated by adding together the two previous numbers in the sequence. Or:

1, 3, 6, 10, 15, 21 . . .

known as the triangular numbers, which counts the number of stones you'd need to build a triangle with an extra layer each time. You'll also find one of the most enigmatic sequences in the mathematical books:

2, 3, 5, 7, 11, 13 . . .

with the explanation that these are the indivisible numbers or prime numbers. This entry doesn't give you a nice formula to generate the next one because that is one of the big open problems mathematicians have not been able to solve. Get an algorithm to crack this sequence successfully and I think we would all pack up and go home. The database includes some of the sequences my own research is obsessed with, including sequence number 158079, which begins:

1, 2, 5, 15, 67, 504, 9310 . . .

These numbers count the number of symmetrical objects with 3, 3^2, 3^3, 3^4, 3^5, 3^6, 3^7 symmetries. My research has shown that they follow a Fibonacci-like rule, but I am still on the search for what particular combination of previous numbers in the sequence you need in order to get the next number.

Colton decided he would get his algorithm to try to identify new sequences and to explain why they might be interesting. Among its candidates is a sequence Colton's colleague Toby Walsh named 'refactorable numbers'. These are numbers which are such that the number of divisors is itself a divisor (so, 9 is refactorable, because this has three divisors, and 3 divides 9). It's a rather bizarre-sounding number but the algorithm did conjecture that all odd refactorable numbers would be perfect squares. Although it couldn't prove this, the suggestion was enough to intrigue Colton, who proved that this was in fact true, leading to the publication of a journal paper explaining the proof. It transpired that while the sequence was missing from the Encyclopaedia, refactorable numbers had already been invented, although none of the algorithmic conjectures about them had been made. Could this be the first hints of a Great Automatic Mathematizator appearing over the horizon?

Have AI got news for you

Where writing algorithms are coming into their own is in translating undigested data into news stories. Every week companies across the world are releasing data about their earnings. In the past a news organisation like Associated Press would have to assign a phalanx of journalists to plough through the data and then compile a report on how the companies were faring. It was boring and inefficient. You could probably cover about 1000 companies during a year, but that meant so many other companies that people might be interested in were not reported on. Journalists in the office dreaded being chosen to write these stories. They were the bane of any reporter's existence.

So there are few journalists crying over Associated Press's decision to enlist machines to help tell these stories. Algorithms like Wordsmith, created by Automated Insights, or Narrative Science's Quill are now helping to churn out data-driven stories that match the dry efficiency of many of the articles that humans used to have to produce for the Associated Press. Most times you will know only when you come to the bottom of the article that a machine wrote the piece. The algorithms are freeing the journalists to write about the bigger picture.

Data-mining algorithms are also increasingly useful to businesses producing the numbers reported on at Associated Press. An algorithm can take huge swathes of business information and turn unreadable spreadsheets into stories written in a language that company employees can understand. It can pick out subtle changes from month to month in the manufacturing output of a company or turn data about employee work rates into predictions that although John is the most productive employee this month, based on current outputs Susan should be outperforming John by the end of next month. This kind of granular detail could easily be hidden in the spreadsheets and bar charts. When translated into natural language, it becomes a story that resonates. These narratives are becoming particularly important for investors seeking to navigate potential changes in a company's valuation.

But the algorithms are just as at home producing the sort of opinionated snark-laden sports stories that we enjoy reading on the back page of the tabloid newspapers or politically biased stories that will suit the reader's tastes given what they've read to date. Local newspapers that have few reporters can't hope to cover all the local sport, so increasingly they are using algorithms to change football or baseball results into readable news stories. Some journalists of course have been horrified by the idea of their job being done by a machine and started trying to call out reports as clearly written by an algorithm. In one instance a report on the George Washington University sports website had failed to celebrate the remarkable achievement of the pitcher of the opposition team to pitch a perfect game – pitching to twenty-seven batsmen over nine innings and ensuring that none of them got even to first base. The journalists declared that this was the sort of rare event that an algorithm could never be programmed to report on.

It turned out the article was actually written by a human who probably supported the home baseball team that had suffered the humiliating defeat and had buried the achievement in the penultimate paragraph. The team at Narrative Science were interested to take the data from the game to see what their algorithm would make of it. Here is the beginning of the article generated just from the numerical data it was given:

> Tuesday was a great day for W. Roberts, as the junior pitcher threw a perfect game to carry Virginia to a 2–0 victory over George Washington at Davenport Field.
>
> Twenty-seven Colonials came to the plate and the Virginia pitcher vanquished them all, pitching a perfect game. He struck out 10 batters while recording his momentous feat. Roberts got Ryan Thomas to ground out for the final out of the game.

Algorithms 1 Human journalist 0.

As well as real-life sports events, increasingly people are more interested in the fantasy teams they have put together. There are nearly 60 million people in the US and Canada who have put together fictional combinations of players from the NFL to make teams to compete with their friends, spending on average twenty-nine hours a year managing their team. Yahoo have started using Wordsmith to produce personalised news stories about the fictional teams from the NFL data generated each week. There is no way that humans could produce the millions of news stories that are sent out each week to sate the appetite of players to find out how their teams are doing.

Of course, there is a sinister side to algorithms telling us the news. A story is a powerful political tool, as history repeatedly reminds us. Recent research has taught us how little data and evidence will change people's minds. It is only when this data and evidence is woven into a story that it has the power to persuade and change minds. Someone who is convinced it is dangerous to vaccinate their child will rarely be persuaded by statistics on the power of vaccines to stop the spread of disease. But tell them a story about someone who has come down with measles or smallpox and combine that story with the data and you stand a chance of getting them to reconsider. As George Monbiot put it in *Out of the Wreckage*: 'The only thing that can displace a story is a story.'

The fact that stories can be used to change opinions is something companies like Cambridge Analytica have exploited ruthlessly. By harvesting the personal information of 87 million Facebook users with an app called 'This is your digital life', Cambridge Analytica was able to draw up psychological profiles that could then be matched with news stories to influence the way people might vote. The algorithms started by randomly assigning stories, but they gradually learned which ones attracted clicks.

They soon picked up that young, conservatively minded whites in the US responded positively to phrases like 'drain the swamp' or to the idea of building a wall to keep out immigrants. So the algorithm started filling their Facebook pages with algorithmically generated stories to feed their appetite for swamps and walls. It ensured that these stories were put in front of the people whose views were most likely to be changed by them and not wasted on those who were more likely to be unaffected.

When the news broke that Cambridge Analytica had effectively manipulated the electorate, the backlash brought the company down – ironically revealing exactly what it had banked on: the power of a news story to influence events.

While Cambridge Analytica may have folded, there are many other companies out there that continue to mine data to squeeze out strategic advantages for those willing to pay. If we want to retain a modicum of control over our lives, it is important that we understand how our emotions and political opinions are being pushed and pulled around by these algorithms, and how, given the same information, each one will spin its own particular yarn, tailored to fan our hang-ups and views.

I should come clean at this point and admit that I didn't write all of this book myself. I succumbed to the offer made by a modern-day version of Roald Dahl's Great Automatic Grammatizator. A 350-word section of the book was written by an algorithm that specialises in producing short-form essays based on a number of key words that you feed in. Did it pass the literary Turing Test? Did you notice?

One of the dangers of allowing any algorithm to write articles based on existing texts is of course plagiarism. The algorithm could get me into trouble. I managed to chase it back through the web and found an article on another website with some remarkable similarities to the paragraphs I'd been offered. I guess when I get sued for plagiarism by the author of that article I'll know that AI-generated text isn't all it's cracked up to be.

For all of its variability and innovation, the current state of algorithmic storytelling is not a threat to authors. The Great Automatic Grammatizator remains a human fantasy. Even the logical stories we mathematicians tell one another are the preserve of the human mind. There are so many stories to tell that choosing which ones are worth telling remains a challenge. Only a human creator will understand why another human mind would want to follow them on their creative journey. No doubt computers will assist us on our journey, but they will be the telescopes and typewriters, not the storytellers.

16

WHY WE CREATE: A MEETING OF MINDS

Creativity is the essence of that which is not mechanical.
Yet every creative act is mechanical – it has its
explanation no less than a case of the hiccups does.

Douglas Hofstadter

Computers are a powerful new tool in extending the human code. We have discovered new moves in the game of Go that have expanded the way we play. Jazz musicians have heard parts of their sound world that they never realised were part of their repertoire. Mathematical theorems that were impossible for the human mind to navigate are now within reach. Adversarial algorithms are creating art that rivals work shown at international art fairs. And yet my journey has not produced anything that presents an existential threat to what it means to be a creative human. Not yet at least.

I think throughout my journey I've fluctuated between being absolutely convinced that an algorithm will never get anywhere near what it is humans are doing when they paint, compose or write. And yet I'll come back to the realisation that all the decisions that are being made by an artist are being driven at some level by an algorithmic response of the body to the world around it. How easy will it be for a machine to have a response as rich and as complex as the one the human code produces? The human code has evolved over millions of years. The question is how fast could that evolution be speeded up?

The new ideas of machine learning I think challenge many of the traditional arguments about why machines can never be creative. Machine learning does not require the programmer to understand how Bach composed his chorales, because the algorithm can take the data and learn for itself. And such learning is leading to new insights into the creative process of human artists. There is a challenge that such a process of creativity will only produce more of the same. How can it break out of the data that it is learning on? But even here we have seen the possibility to discover new unexplored regions of an artist's world. The jazz musician recognises the output of the algorithm as part of his sound world and yet the result is a new way to combine his riffs.

Many will concede that exploratory creativity and combinational creativity could be something that an algorithm can achieve because it is relying on previous creativity by humans that the algorithm then extends or combines. But the challenge of algorithmically producing transformational creativity is one that traditionally seems impossible. How can an algorithm that is stuck inside a system find a way to break out and do something that shocks us? But here again the new approach to AI sees how we can create meta-algorithms that are encouraged to break the rules and see what happens. Transformational creativity is not really *ex nihilo* but a perturbation of existing systems.

What about the challenge that this is still all the creation of the coder? Scientists are beginning to recognise that genuinely new things can emerge out of combinations of old things. That the whole can be more than the sum of its parts. The idea of an emergent phenomenon has a lot of cachet in science at the moment. It is an antidote to the reductionist view, whereby everything can be boiled down to atoms and equations. Consciousness or the wetness of water are both heralded as emergent phenomena. One molecule of H_2O is not wet; only a collection of molecules at some point has the property of wetness. One neuron is not conscious; many can be. There is some interesting speculation that time is not absolute and that it emerges as a consequence of humans' incomplete knowledge of the universe.

It may be that we should regard products of our new complex algorithms a bit like emergent phenomena. Yes, they are a consequence of the rules that created them, but they are still more than the sum of their parts. Some artists, novelists especially, say that once they start a project it's as if the process takes on a life of its own. William Golding talked about how his stories seemed to become independent of him: 'The author becomes a spectator, appalled or delighted, but a spectator.' Is a similar disconnect between coder and code the goal if we are to prove Lovelace wrong?

Another broadside fired at AI creativity is the fact that it can't reflect on its own output and make a judgement about whether it is good or bad, worth sharing or deleting. But this ability to self-reflect too has been shown to be possible. One can create adversarial algorithms that can judge whether a piece of art is too derivative or has strayed outside the boundaries of what we consider art. So why do I still feel that anything to match human creativity is still way beyond the reach even of these amazing new tools?

At the moment all the creativity in machines is being initiated and driven by the human code. We are not seeing machines compelled to express themselves. They don't really seem to have anything to say beyond what we are getting them to do. They are the ventriloquist's dummy providing the mouthpiece for our urge still to express ourselves. And that creative urge is an expression of our belief in free will. We can live our lives like automata or we can suddenly make the choice to stop and break out of the routine and create something that is new. Our creativity is intimately bound up with our free will, something that it seems impossible to automate. To program free will would be to contradict what free will means. Although, then again, we might end up asking whether our free will is an illusion which just masks the complexity of our underlying algorithmic process.

The current drive by humans to create algorithmic creativity is in the most part not one fuelled by the desire for extending artistic creation but rather enlarging a company bank balance. There is a huge amount of hype about AI. There are too many initiatives that are branded as AI but which are little more than statistics or data science. Just as any company wishing to make it at the turn of the millennium would put .com on the end of its name, today it is the addition of the tag AI or Deep which is what companies are using to jump on the bandwagon.

Companies would love to be able to convince an audience that this AI is so great that it can write articles on its own, that it can compose music, paint Rembrandts. It is all fuel for convincing customers that the AI on offer is going to transform their business too if they invest. But when you look beyond the hype you see it is still the human code that is driving this revolution.

It is interesting to go back to the origins of our obsession with creativity. Creativity as meaning something novel with value is actually a very twentieth-century capitalist take on the word. It has its origins in the self-help books written by the advertising executive Alex Osborn in the 1940s. Books like *Your Creative Power* and *Brainstorming* were looking to realise creativity in individuals and organisations. But before this rather commercial attitude towards valuable novelty, creative activity was meant to capture humans' attempts to understand being in the world.

We can continue along as automata without acting at all in the world or we can choose to break out of those constraints to understand our place in it. As the psychologist Carl Rogers expresses it in his essay 'Towards a Theory of Creativity', it is 'the urge to expand, extend, develop, mature – the tendency to express and activate all the capacities of the organism, to the extent that such activation enhances the organism or the self'. Creativity is about humans asserting they are not machines. Although today's AI is a long way from matching human creativity, it has its part to play in making us more creative. Strangely it might end up helping humans to behave less mechanically by giving us the creative spark that we are so often missing in our daily lives.

Ultimately I think the word 'self' in Rogers' analysis is key here. For me human creativity and consciousness are inextricably linked. I don't think we can understand why we are creative without the concept of consciousness. Although it is impossible to establish, I would suspect that the two emerged at the same time in our species. With the realisation of our own inner world came the desire to know ourselves and share it with others who cannot directly access the self of the organism driven to create. For the Brazilian writer Paulo Coelho the urge is part of what it means to be human: 'Writing means sharing. It's part of the human condition to want to share things – thoughts, ideas, opinions.' For Jackson Pollock: 'Painting is self-discovery. Every good artist paints what he is.' One of the challenges of consciousness is that it is impossible for me to feel what it means to be you. Is your pain anything like mine? Is the ecstasy you feel at a moment of extreme joy the same feeling I have? This is something science will never be able to answer. A story or painting is better than any fMRI that might try to scan our emotional state. Our outpourings of creative art, music and literature may be the better canvas to explore what it means to be a conscious emotional human being.

'The greatest benefit we owe to the artist, whether painter, poet or novelist, is the extension of our sympathies . . . Art is the nearest thing to life; it is a mode of amplifying experience and extending our contact with our fellow-men beyond the bounds of our personal lot,' wrote the novelist George Eliot.

The political role of art in mediating an individual's engagement with the group is also key. It is often about the desire to change the status quo: to break humanity out of following the current rules of the game; to create a better place, or maybe just a different place, for our fellow humans. This was certainly a motivation for George Orwell: 'When I sit down to write a book, I do not say to myself, "I am going to produce a work of art." I write it because there is some lie that I want to expose, some fact to which I want to draw attention, and my initial concern is to get a hearing.' For Zadie Smith there is a political motivation to her storytelling: 'Writing is my way of expressing – and thereby eliminating – all the various ways we can be wrong-headed.'

Why do people become audiences for these artistic outputs? Perhaps it's in part an act of creativity that the audience member can take part in themselves. It often requires some creativity to be able to engage with many works of art that deliberately leave room for the viewer or the reader or the listener to bring their story to bear. Ambiguity is an important part of artistic creation because this is where the audience can be creative.

There is some argument that our whole lives are an act of creativity. Shakespeare was one of the first to recognise this with his famous speech from *As You Like It*:

All the world's a stage
And all the men and women merely players;
They have their exits and their entrances;
And one man in his time plays many parts

The American psychologist Jerome Bruner believed 'a self is probably the most impressive work of art we ever produce, surely the most intricate.' The works that we call art, whether music, paintings or poems, are almost like the by-products or the breaking off of a piece of this act of creation of the self. Again we come back to the lack of self in a machine being a fundamental barrier to creativity.

Creativity is very tied up with mortality, something very much coded into what it means to be human. Many in search of meaning for their existence who find the religious stories meaningless will perhaps look to leave something behind them that will outlast their finite existence, whether it be a painting, a novel, a theorem, a child. Are these all attempts to use creativity as an act of cheating death?

And maybe death is part of why we value acts of creativity. If Cope was successful in producing an algorithm that could churn out endless Chopin mazurkas such that it's as if it has made Chopin immortal, then does that make us happy? I don't think so. It would begin to devalue the pieces that Chopin did compose. Isn't it like the Library of Babel, where because it contains everything it ends up containing nothing? It is the choices that Chopin made that are important. Hasn't the game of chess in part been devalued by the power of the computer just to churn out wins?

Perhaps the human battle with chess, music, mathematics, painting, is part of where the value comes from. If we could ultimately solve death and create immortal versions of ourselves, many believe that would devalue life, make each day meaningless. It is our mortality which somehow matters. Being aware of our mortality is one of the costs of consciousness. My iPhone does not yet realise that it is going to be obsolete in two years' time. But when it is aware, will it be driven to try to leave something behind that provides proof of its existence?

Until a machine has become conscious I don't think it will be any more than a tool for extending human creativity. Do we have any idea of what it will take to create consciousness in a machine? There is some research about the difference between the network of the human brain when it is awake and when it is in deep stage 4 sleep, our most unconscious state. The key seems to be a certain feedback quality. In the awake conscious brain we see activity start at one place in the brain and cascade across the network, followed by a feeding back to the original source, a sequence which is then repeated over and over as if the feedback were updating our experience. In the sleeping brain we just see very localised behaviour with no such feedback. The machine learning that has seen AI go from successive winters to sudden heatwave has a certain quality of this feedback behaviour, of learning from its interactions. Could we be making the first steps towards AI that might ultimately become conscious and then truly creative?

But what if a machine does become conscious? How could we ever know? Would its consciousness be anything like ours? I don't believe there is any fundamental reason why at some point in the future we can't make a machine that is conscious. I think it will need to tap into all the sciences to do that. And once we are successful I expect that machine consciousness will be very different from our own. And I'm sure it will want to tell us what it's like. It's then that the creative arts will be key to giving each other access to what it feels like to be the other.

Storytelling rather than an fMRI scanner might be our best way of trying to get some hold on what it feels like to be my iPhone. That's why of all the efforts that have so far emerged from the field of literary creativity it is *The Seeker* which feels closest to what we might expect to see from a conscious machine: an algorithm trying to empathise with humans and understand our world. Could this be why storytelling might be an important tool as we move into the future and begin to wonder whether our technology might one day become conscious? Surely that will be the reason why a computer might feel compelled to tell stories rather than that compulsion coming from us?

Just as story is a powerful political tool for binding human society, if machines become conscious, then the ability to share stories might save us from the horrors of the world of AI often depicted in our scenarios of a future with machines. It is striking to recall the novelist Ian McEwan's response to the horrors of the attack of 9/11 in America and his appeal to the importance of empathy in moving forward:

> If the hijackers had been able to imagine themselves
> into the thoughts and feelings of the passengers, they
> would have been unable to proceed. It is hard to be
> cruel once you permit yourself to enter the mind of your
> victim. Imagining what it is like to be someone other than
> yourself is at the core of our humanity. It is the essence
> of compassion, and it is the beginning of morality.

Being able to share our conscious worlds through stories is what makes us human. No other species is likely to do anything like this. If machines become conscious then instilling empathy in the machine might save us from the Terminator story we've concocted for a possible future with machines.

Riedl, the lead researcher on the storytelling machine Scheherazade-IF, was quite struck by how the algorithm didn't choose strange inhuman paths through the set of alternatives it had generated. It learned from the way humans told the stories: 'We have recently been able to show that AI trained on stories cannot behave psychotically, except under the most extreme circumstances. Thus, computational narrative intelligence could alleviate concerns about renegade "evil AI" taking over the earth.'

If and when the singularity strikes, humanity's fate will depend on a mutual understanding with conscious machines. But, as Wittgenstein said, if a lion could talk we probably wouldn't understand it. The same applies to machines. If they become conscious, it's unlikely to be something that humans will initially understand. Ultimately it will be their paintings, their music, their novels, their creative output, even their mathematics that will give us any chance to crack the machine's code and feel what it's like to be a machine.

ILLUSTRATIONS

p. 197 The Alberti bass pattern, from Mozart's piano sonata in C, K545.

p. 199 David Cope's analysis of Scriabin's Prelude No.1, Op. 16.

p. 222 Mihaly Csikszentmihalyi's theory of flow. Oliverbeatson / Wikimedia Commons / Public Domain.

p. 250 Diagram demonstrating that if you add together N consecutive odd numbers, you will get the Nth square number.

FURTHER READING

Machine Learning: The Power and Promise of Computers That Learn by Example. The report by the Royal Society that Margaret Boden, Demis Hassabis and I helped prepare. Issued in April 2017. It can be viewed online at http://royalsociety.org/machine-learning.

Books

Alpaydin, Ethem, *Machine Learning*, MIT Press, 2016

Barthes, Roland, *S/Z*, Farrar, Straus and Giroux, 1991

Berger, John, *Ways of Seeing*, Penguin Books, 1972

Bishop, Christopher, *Pattern Recognition and Machine Learning*, Springer, 2007

Boden, Margaret, *The Creative Mind: Myths and Mechanisms*, Weidenfeld and Nicolson, 1990

——, *AI: Its Nature and Future*, OUP, 2016

Bohm, David, *On Creativity*, Routledge, 1996

Bostrom, Nick, *Superintelligence: Paths, Dangers, Strategies*, OUP, 2014

Braidotti, Rosi, *The Posthuman*, Polity Press, 2013

Brandt, Anthony and David Eagleman, *The Runaway Species: How Human Creativity Remakes the World*, Canongate, 2017

Brynjolfsson, Erik and Andrew McAfee, *The Second Machine Age: Work, Progress, and Prosperity in a Time of Brilliant Technologies*, Norton, 2014

Cawelti, John, *Adventure, Mystery, and Romance: Formula Stories as Art and Popular Culture*, University of Chicago Press, 1977

Cheng, Ian, *Emissaries Guide to Worlding*, Verlag der Buchhandlung Walther Konig, 2018; Serpentine Galleries/ Fondazione Sandretto Re Rebaudengo, 2018

Cope, David, *Virtual Music: Computer Synthesis of Musical Style*, MIT Press, 2001

——, *Computer Models of Musical Creativity*, MIT Press, 2005

Domingos, Pedro, *The Master Algorithm: How the Quest for the Ultimate Learning Machine Will Remake Our World*, Basic Books, 2015

Dormehl, Luke, *The Formula: How Algorithms Solve All Our Problems . . . and Create More*, Penguin Books, 2014

——, *Thinking Machines: The Inside Story of Artificial Intelligence and Our Race to Build the Future*, W. H. Allen, 2016

Eagleton, Terry, *The Ideology of the Aesthetic*, Blackwell, 1990

Ford, Martin, *The Rise of the Robots: Technology and the Threat of Mass Unemployment*, Oneworld, 2015

Fuentes, Agustín, *The Creative Spark: How Imagination Made Humans Exceptional*, Dutton, 2017

Gaines, James, *Evening in the Palace of Reason: Bach Meets Frederick the Great in the Age of Enlightenment*, Fourth Estate, 2005

Ganesalingam, Mohan, *The Language of Mathematics: A Linguistic and Philosophical Investigation*, Springer, 2013

Gaut, Berys and Matthew Kieran (eds.), *Creativity and Philosophy*, Routledge, 2018

Goodfellow, Ian, Yoshua Bengio and Aaron Courville, *Deep Learning*, MIT Press, 2016

Harari, Yuval Noah, *Homo Deus: A Brief History of Tomorrow*, Harvill Secker, 2016

Hardy, G. H., *A Mathematician's Apology*, CUP, 1940

Harel, David, *Computers Ltd: What They Really Can't Do*, OUP, 2000

Hayles, N. Katherine, *Unthought: The Power of the Cognitive Nonconscious*, University of Chicago Press, 2017

Hofstadter, Douglas, *Gödel, Escher, Bach: An Eternal Golden Braid*, Penguin Books, 1979

——, *Fluid Concepts and Creative Analogies: Computer Models of the Fundamental Mechanisms of Thought*, Basic Books 1995

——, *I am a Strange Loop*, Basic Books, 2007

Kasparov, Garry, *Deep Thinking: Where Artificial Intelligence Ends and Human Creativity Begins*, John Murray, 2017

McAfee, Andrew and Erik Brynjolfsson, *Machine Platform Crowd: Harnessing Our Digital Future*, Norton, 2017

McCormack, Jon and Mark d'Inverno (eds.), *Computers and Creativity*, Springer, 2012

Monbiot, George, *Out of the Wreckage: A New Politics for an Age of Crisis*, Verso, 2017

Montfort, Nick, *World Clock*, Bad Quarto, 2013

Moretti, Franco, *Graphs, Maps, Trees: Abstract Models for Literary History*, Verso, 2005

Paul, Elliot Samuel and Scott Barry Kaufman (eds.), *The Philosophy of Creativity: New Essays*, OUP, 2014

Shalev-Shwartz, Shai and Shai Ben-David, *Understanding Machine Learning: From Theory to Algorithms*, CUP, 2014

Steels, Luc, *The Talking Heads Experiment: Origins of Words and Meanings*, Language Science Press, 2015

Steiner, Christopher, *Automate This: How Algorithms Took Over the Markets, Our Jobs, and the World*, Penguin Books, 2012

Tatlow, Ruth, *Bach and the Riddle of the Number Alphabet*, CUP, 1991

——, *Bach's Numbers: Compositional Proportions and Significance*, CUP, 2015

Tegmark, Max, *Life 3.0: Being Human in the Age of Artificial Intelligence*, Allen Lane, 2017

Wilson, Edward O., *The Origins of Creativity*, Allen Lane, 2017

Yorke, John, *Into the Woods: A Five Act Journey into Story*, Penguin Books, 2013

Papers

For papers with references to arXiv visit the open access archive of papers at https://arxiv.org/.

Alemi, Alex A., et al., 'DeepMath: Deep Sequence Models for Premise Selection', arXiv:1606.04442v2 (2017)

Athalye, Anish, et al., 'Synthesizing Robust Adversarial Examples', in *Proceedings of the 35th International Conference on Machine Learning*, arXiv:1707.07937v3 (2018)

Bancerek, Grzegorz, et al., 'Mizar: State-of-the-Art and Beyond', in *Intelligent Computer Mathematics*, pp. 261–79, Springer, 2015

Barbieri, Francesco, Horacio Saggion and Francesco Ronzano, 'Modelling Sarcasm in Twitter: A Novel Approach', *WASSA@ ACL* (2014)

Bellemare, Marc, et al., 'Unifying Count-Based Exploration and Intrinsic Motivation', in *Advances in Neural Information Processing Systems*, pp. 1471–9, NIPS Proceedings, 2016

Bokde, Dheeraj, Sheetal Girase and Debajyoti Mukhopadhyay, 'Matrix Factorization Model in Collaborative Filtering Algorithms: A Survey', *Procedia Computer Science*, vol. 49, 136–46 (2015)

Briot, Jean-Pierre and François Pachet, 'Music Generation by Deep Learning: Challenges and Directions', arXiv:1712. 04371 (2017)

Briot, Jean-Pierre, Gaëtan Hadjeres and François Pachet, 'Deep Learning Techniques for Music Generation: A Survey', arXiv:1709.01620 (2017)

Brown, Tom B., et al., 'Adversarial Patch', arXiv:1712.09665 (2017)

Cavallo, Flaminia, Alison Pease, Jeremy Gow and Simon Colton, 'Using Theory Formation Techniques for the Invention of Fictional Concepts', *in Proceedings of the Fourth International Conference on Computational Creativity* (2013)

Clarke, Eric F., 'Imitating and Evaluating Real and Transformed Musical Performances', *Music Perception: An Interdisciplinary Journal*, vol. 10, 317–41 (1993)

Colton, Simon, 'Refactorable Numbers: A Machine Invention', *Journal of Integer Sequences*, vol. 2, article 99.1.2 (1999)

——, 'The Painting Fool: Stories from Building an Automated Painter', in Jon McCormack and Mark d'Inverno (eds.), *Computers and Creativity*, Springer, 2012

—— and Stephen Muggleton, 'Mathematical Applications of Inductive Logic Programming', *Machine Learning*, vol. 64(1), 25–64 (2006)

—— and Dan Ventura, 'You Can't Know My Mind: A Festival of Computational Creativity', in *Proceedings of the Fifth International Conference on Computational Creativity* (2014)

——, et al., 'The "Beyond the Fence" Musical and "Computer Says Show" Documentary', in *Proceedings of the Seventh International Conference on Computational Creativity* (2016)

d'Inverno, Mark and Arthur Still, 'A History of Creativity for Future AI Research', in *Proceedings of the Seventh International Conference on Computational Creativity* (2016)

du Sautoy, Marcus, 'Finitely Generated Groups, p-Adic Analytic Groups and Poincaré Series', *Annals of Mathematics*, vol. 137, 639–70 (1993)

du Sautoy, Marcus, 'Counting Subgroups in Nilpotent Groups and Points on Elliptic Curves', *J. reine angew. Math*, 549, 1–21 (2002)

Ebcioglu, Kemal, 'An Expert System for Harmonizing Chorales in the Style of J. S. Bach', *Journal of Logical Programming*, vol. 8, 145–85 (1990)

Eisenberger, Robert and Justin Aselage, 'Incremental Effects of Reward on Experienced Performance Pressure: Positive Outcomes for Intrinsic Interest and Creativity', *Journal of Organizational Behavior*, 30(1), 95–117 (2009)

Elgammal, Ahmed and Babak Saleh, 'Quantifying Creativity in Art Networks', in *Proceedings of the Sixth International Conference on Computational Creativity* (2015)

—— and ——, 'Large-Scale Classification of Fine-Art Paintings: Learning the Right Metric on the Right Feature', arXiv: 1505.00855 (2015)

Elgammal, Ahmed, et al., 'CAN: Creative Adversarial Networks Generating "Art" by Learning about Styles and Deviating from Style Norms', arXiv:1706.07068 (2017)

Ferrucci, David A., 'Introduction to "This is Watson"', *IBM Journal of Research and Development,* vol. 56(3.4), 1.1–1.15 (2012)

Ganesalingam, Mohan and W. T. Gowers, 'A Fully Automatic Theorem Prover with Human-Style Output', *Journal of Automated Reasoning*, vol. 58(2), 253–91 (2016)

Gatys, Leon A., Alexander S. Ecker and Matthias Bethge, 'A Neural Algorithm of Artistic Style', arXiv:1508.06576 (2015)

Gondek, David, et al., 'A Framework for Merging and Ranking of Answers in DeepQA', *IBM Journal of Research and Development,* vol. 56(3.4), 14:1–14:12 (2012)

Gonthier, Georges, 'A Computer-Checked Proof of the Four Colour Theorem', Microsoft Research Cambridge (2005)

——, 'Formal Proof: The Four-Color Theorem', *Notices of the AMS*, vol. 55, 1382–93 (2008)

——, et al., 'A Machine-Checked Proof of the Odd Order Theorem', *Interactive Theorem Proving, Proceedings of the Fourth International Conference on ITP* (2013)

Goodfellow, Ian J., 'NIPS 2016 Tutorial: Generative Adversarial Networks', arXiv:1701.00160 (2016)

Guzdial, Matthew J., et al., 'Crowdsourcing Open Interactive Narrative', *Tenth International Conference on the Foundations of Digital Games* (2015)

Hadjeres, Gaëtan, François Pachet and Frank Nielsen, 'DeepBach: A Steerable Model for Bach Chorales Generation', arXiv:1612.01010 (2017)

Hales, Thomas, et al., 'A Formal Proof of The Kepler Conjecture', *Forum of Mathematics, Pi*, vol. 5, e2 (2017)

Hermann, Karl Moritz, et al., 'Teaching Machines to Read and Comprehend', in *Advances in Neural Information Processing Systems, NIPS Proceedings* (2015)

Ilyas, Andrew, et al., 'Query-Efficient Black-Box Adversarial Examples', arXiv:1712.07113 (2017)

Khalifa, Ahmed, Gabriella A. B. Barros and Julian Togelius, 'DeepTingle', arXiv:1705.03557 (2017)

Koren, Yehuda, Robert M. Bell and Chris Volinsky, 'Matrix Factorization Techniques for Recommender Systems', *Computer Journal*, vol. 42(8), 30–37 (2009)

Li, Boyang and Mark O. Riedl, 'Scheherazade: Crowd-Powered Interactive Narrative Generation', *29th AAAI Conference on Artificial Intelligence* (2015)

Llano, Maria Teresa, et al., 'What If a Fish Got Drunk? Exploring the Plausibility of Machine-Generated Fictions', in *Proceedings of the Seventh International Conference on Computational Creativity* (2016)

Loos, Sarah, et al., 'Deep Network Guided Proof Search', arXiv: 1701.06972v1 (2017)

Mahendran, Aravindh and Andrea Vedaldi, 'Understanding Deep Image Representations by Inverting Them', *Proceedings of the 2015 IEEE Conference on Computer Vision and Pattern Recognition (CVPR)*, 5188–96 (2015)

Mathewson, Kory Wallace and Piotr W. Mirowski, 'Improvised Comedy as a Turing Test', arXiv:1711.08819 2017

Matuszewski, Roman and Piotr Rudnicki, 'MIZAR: The First 30 Years', *Mechanized Mathematics and Its Applications*, vol. 4, 3–24 (2005)

Melis, Gábor, Chris Dyer and Phil Blunsom, 'On the State of the Art of Evaluation in Neural Language Models', arXiv:1707. 05589v2 (2017)

Mikolov, Tomas, et al., 'Efficient Estimation of Word Representations in Vector Space', arXiv:1301.3781 (2013)

Mnih, Volodymyr, et al., 'Playing Atari with Deep Reinforcement Learning', arXiv:1312.5602v1 (2013)

Mnih, Volodymyr, et al., 'Human-Level Control through Deep Reinforcement Learning', *Nature*, vol. 518(7540), 529–33 (2015)

Narayanan, Arvind and Vitaly Shmatikov, 'Robust De-anonymization of Large Datasets (How to Break Anonymity of the Netflix Prize Dataset)', arXiv:cs/0610105 v2 (2007)

Nguyen, Anh Mai, Jason Yosinski and Jeff Clune, 'Deep Neural Networks are Easily Fooled: High Confidence Predictions for Unrecognizable Images', *Proceedings of the 2015 IEEE Conference on Computer Vision and Pattern Recognition (CVPR)*, 427–36 (2015)

Pachet, François, 'The Continuator: Musical Interaction with Style', presented at the *International Computer Music Conference, Journal of New Music Research*, vol. 31(1) (2002)

—— and Pierre Roy, 'Markov Constraints: Steerable Generation of Markov Sequences', *Constraints*, vol. 16, 148–72 (2011)

——, et al., 'Reflexive Loopers for Solo Musical Improvisation', presented at the ACM SIGCHI Conference on Human Factors in Computing Systems (2013)

Riedl, Mark O. and Vadim Bulitko, 'Interactive Narrative: An Intelligent Systems Approach', *AI Magazine*, vol. 34, 67–77 (2013)

Roy, Pierre, Alexandre Papadopoulos and François Pachet, 'Sampling Variations of Lead Sheets', arXiv:1703.00760 (2017)

Silver, David, et al., 'Mastering the Game of Go with Deep Neural Networks and Tree Search', *Nature*, vol. 529(7587), 484–9 (2016)

Stern, David H., Ralf Herbrich and Thore Graepel, 'Matchbox: Large Scale Online Bayesian Recommendations', in *WWW '09: Proceedings of the 18th International World Wide Web Conference*, 111–20 (2009)

Tesauro, Gerald, et al., 'Analysis of Watson's Strategies for Playing Jeopardy!' *Journal of Artificial Intelligence Research*, vol. 47(1), 205–51 (2014)

Torresani, Lorenzo, Martin Szummer and Andrew Fitzgibbon, 'Efficient Object Category Recognition Using Classemes', in *Computer Vision: ECCV 2010*, Springer, 2010, pp. 776–89

Wang, C., et al., 'Relation Extraction and Scoring in DeepQA', *IBM Journal of Research and Development,* vol. 56(3:4), 9:1–9:12 (2012)

Weiss, Ron J., et al., 'Sequence-to-Sequence Models Can Directly Translate Foreign Speech', *INTERSPEECH 2017*, 2625–9 (2017)

Yu, Lei, et al., 'Deep Learning for Answer Sentence Selection', arXiv:1412.1632v1 (2014)

Zeilberger, Doron, 'What is Mathematics and What Should It Be?', arXiv:1704.05560v1 (2017)

Courses

Eremenko, Kirill, Hadelin de Ponteves and the SuperDataScience Team, 'Machine Learning A–Z', Udemy

Fiebrink, Rebecca, 'Machine Learning for Musicians and Artists', Goldsmiths, University of London via Kadenze

Hinton, Geoffrey, 'Neural Networks for Machine Learning', University of Toronto via Coursera

Irizarry, Rafael, 'Data Science: Machine Learning', Harvard University via edX

Ng, Andrew, 'Machine Learning', Stanford University via Coursera

Paisley, John, 'Machine Learning', Columbia University via edX

Extras

To find out how many websites there are on the internet today: http://www.internetlivestats.com/.

Great resource for AlphaGo including all the games with Lee Sedol: https://deepmind.com/research/alphago/.

To read *The Seeker* by thricedotted, see the following link at github, the hosting service for computer code: https://github.com/thricedotted/theseeker.

Other novels created as part of NaNoGenMo can be found at:
https://nanogenmo.github.io/.

Loebner prize transcripts can be found at:
https://www.aisb.org.uk/events/loebner-prize.

To view works by and articles about AARON:
http://aarons home.com/.

To view works by and articles about 'The Painting Fool':
http://www.thepaintingfool.com/.

To see the paintings generated by the Elgammal's Creative
Adversarial Network: https://sites.google.com/site/digi
humanlab/home.

To see how good algorithms are at recognising an image:
https://cloud.google.com/vision/.

To listen to 'Experiments in Musical Intelligence' by David Cope:
http://artsites.ucsc.edu/faculty/cope/.

For work produced by Botnik: http://botnik.org/.

To see the results of Ganesalingam and Gowers' computer proof
experiment: https://gowers.wordpress.com/.

'The Next Rembrandt' can be viewed here:
https://www.next rembrandt.com/.

ACKNOWLEDGEMENTS

Many thanks to all the people and algorithms I have encountered over the years who have made this book possible. I'm especially grateful to the Royal Society for asking me to serve on the Machine Learning Committee. I usually dread committees, but these were meetings I always enjoyed attending. The following humans played a crucial role in making this book a reality.

My editors: Louise Haines at 4th Estate and Joy de Menil at
  Harvard University Press.
My agent: Antony Topping at Greene & Heaton.
My assistant editor: Sarah Thickett at 4th Estate.
My research assistant: Ben Leigh.
My copy editor: Mark Handsley.
My patron: Charles Simonyi.
My family: Shani, Tomer, Magaly and Ina.

INDEX